Cognitive Analytics
and Reinforcement Learning

Scrivener Publishing
100 Cummings Center, Suite 541J
Beverly, MA 01915-6106

Publishers at Scrivener
Martin Scrivener (martin@scrivenerpublishing.com)
Phillip Carmical (pcarmical@scrivenerpublishing.com)

Cognitive Analytics and Reinforcement Learning

Theories, Techniques and Applications

Edited by

Elakkiya, R.

*Department of Computer Science, Birla Institute of Technology & Science Pilani,
Dubai Campus, UAE*

and

Subramaniyaswamy V.

School of Computing, SASTRA Deemed University, Thanjavur, India

Scrivener
Publishing

This edition first published 2024 by John Wiley & Sons, Inc., 111 River Street, Hoboken, NJ 07030, USA
and Scrivener Publishing LLC, 100 Cummings Center, Suite 541J, Beverly, MA 01915, USA
© 2024 Scrivener Publishing LLC
For more information about Scrivener publications please visit www.scrivenerpublishing.com.

Wiley Global Headquarters
111 River Street, Hoboken, NJ 07030, USA

For details of our global editorial offices, customer services, and more information about Wiley products visit us at www.wiley.com.

Limit of Liability/Disclaimer of Warranty

Library of Congress Cataloging-in-Publication Data

ISBN 978-1-394-21403-7

Cover image: Pixabay.Com
Cover design by Russell Richardson

Set in size of 11pt and Minion Pro by Manila Typesetting Company, Makati, Philippines

Printed in the USA

10 9 8 7 6 5 4 3 2 1

Contents

Preface

Cognitive analytics and reinforcement learning is a transformational force in the field of modern technological breakthroughs, reshaping the decision-making, problem-solving, and innovation landscape. This book offers a means of examining the profound overlap between these two fields and illuminating its significant consequences for business, academia, and research.

The harmonious combination of cognitive analytics and reinforcement learning has emerged as a beacon of wise decision-making in a world that is continually evolving and where data-driven insights drive progress. The symbiotic relationship between cognitive capacities and reinforcement learning techniques has become a cornerstone of dealing with complexity as the globe struggles with complicated problems that call for immediate solutions.

This book's main goal is to shed light on the development of cognitive-enhanced reinforcement learning and all of its numerous uses. Cognitive insights and reinforcement learning dynamics are combined to create a dynamic framework that empowers academics, researchers, and business leaders looking for practical answers to challenging decision-making problems.

This book explores the ideas, methods, and real-world applications that influence the development of cognitive analytics and reinforcement learning. Each chapter contains a narrative of innovation, ranging from the improvement of prediction models to optimizing resource allocation, from identifying healthcare problems to transforming smart cities.

The editors have created a compendium that perfectly captures the essence of reinforcement learning and cognitive analytics—where academic concepts meet practical implementations. This book is a monument to the teamwork of perceptive authors who illuminate the points where different domains converge and provide readers with a broad perspective on the potential that emerges when cognition and reinforcement come together.

We are exceedingly grateful to the authors for their outstanding contribution of knowledge and intelligence, which have made this book a treasure trove of wisdom and innovation. We also thank the readers for joining us on this trip, whether they are seasoned professionals, inquisitive researchers, or enthusiastic learners.

We sincerely hope that this book will spark knowledge, inquiry, and creativity—a physical manifestation of the motivation that propels us ahead in the field of intelligent decision-making.

The Editors
December 2023

Part I

COGNITIVE ANALYTICS IN CONTINUAL LEARNING

Cognitive Analytics in Continual Learning: A New Frontier in Machine Learning Research

Renuga Devi T.[1], Muthukumar K.[2*], Sujatha M.[1†] and Ezhilarasie R.[1]

School of Computing, SASTRA Deemed University, Thanjavur, India
School of Electrical & Electronics Engineering, SASTRA Deemed University, Thanjavur, India

Abstract

The cognitive system that started with automation has now set its benchmark to reach human-centric intelligence. The slow adoption of cognitive systems is most likely due to its meticulous training process. With cognitive computing as its backbone nowadays, any data can be converted into an asset anytime and anywhere. The complexity of data and its abandonment nature demand the coexistence of many technologies to provide deep insights in a domain. A generic artificial intelligence system built on deep learning and natural language processing evolves into a personalized business partner and a life companion that continuously learns. Combining tremendous power, humanity's relationship with technology has undergone incredible shifts. The adaptation and embracement have led to a higher level of intelligence augmentation, mainly in decision support and engagement systems, penetrating its need in various fields, especially in the healthcare industry, business-to-business, industrial marketing, autonomous driving, financial services, manufacturing sectors, and as a human assistant in day-to-day activities. The expensive and complex process of using cognitive systems to get complete resolutions for specific business segments on historical static data and dynamic real-time data should be addressed with Hadoop, Spark, NoSQL, and other technologies that are part of cognitive systems besides NLP, AI, and ML. This chapter begins with an understanding of different analytics and the need of the hour, then gradually penetrates to give insights into cognitive systems, design principles, and

---

Corresponding author: sujatha@cse.sastra.edu
Corresponding author: kmuthukumar@eee.sastra.edu

Elakkiya, R. and Subramaniyaswamy V. (eds.) Cognitive Analytics and Reinforcement Learning: Theories, Techniques and Applications, (3–28) © 2024 Scrivener Publishing LLC

key characteristics of the system, dwelling in the backbone of cognitive systems and its different learning approaches with some prominent use cases.

Keywords: Cognitive computing, machine learning algorithms, natural language processing, artificial intelligence, cognitive analytics

1.1 Introduction

The cognitive age is a continuous trend of massive technological development. The driving force behind this trend is the developing field of cognitive technology, which consists of profoundly disruptive systems that interpret unstructured data, reason to generate hypotheses, learn from experience, and organically interact with humans. With this technology, the capacity to generate insight from all types of data will be critical to success in the cognitive age.

Cognitive computing is likely most notable for upending the conventional IT view that a technology's worth reduces with time; because cognitive systems improve as they learn, they actually grow more useful. This trait makes cognitive technology very valuable for business, and many early adopters are capitalizing on the competitive edge it provides. The cognitive era has arrived, not just because technology has matured, but also because the phenomena of big data necessitate it. The goal of cognitive computing is to be able to solve some uncertain real-world issues comparable to those addressed by the human brain [1].

Since its inception in the 1950s, cognitive science has grown at a rapid pace. Furthermore, as a key component of cognitive science, cognitive computing has a significant influence on artificial intelligence and information technology [2]. Computing systems in the past could gather, transport, and store unstructured data, but they could not interpret it. Cognitive computing systems are intended to foster a better "symbiotic relationship" between humans and technology by replicating human reasoning and problem-solving. Cognitive computing simulates the human brain using computerized models. It is accomplished by the combination of the Von Neumann paradigm and neuromorphic computing, which combines analytic, iterative processes with extremely sophisticated logical and reasoning operations in a very short period of time while utilizing very little power.

The excitement around AI equipment has been dubbed a "renaissance of equipment," as vendors race to manufacture space-explicit or exceptional job-at-hand explicit designs that can fundamentally scale and increase computing productivity [3]. Cognitive systems are probabilistic in nature that hold the capability to adapt and sense the unpredictability and complexity of unstructured input. They analyze that information, organize it, and explain what it means, as well as the reasons for their judgments [4]. Cognitive computing refers to technological platforms that combine reasoning, machine learning, natural language processing, vision, voice, and human computer interaction that replicates the human brain operation and aid in decision-making. The progression of cognitive thought evolves from pure descriptivism through past prediction to prescriptiveness, reflecting a journey from understanding to anticipation and active guidance.

1.2   Evolution of Data Analytics

As we go forward, the graph in Figure 1.1 shows us the benefits that each type of analytics provides.

a) Descriptive Analytics
Acquiring and evaluating facts to explain what has happened. The majority of business reports are descriptive in nature, which is capable of providing historical data summary or explaining differences from one another. Insights from past data are provided in detail by descriptive analytics via data aggregation and data mining but fail to explain the reason behind the insights.

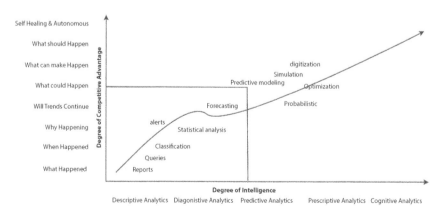

Figure 1.1 Benefits of analytics (source: https://swifterm.com/the-difference-between-descriptive-diagnostic-predictive-and-cognitive-analytics/).

b) Diagnostic Analytics

Diagnostic analytics addresses the reason behind the inference and discovers answers to why questions. The data are compared with past data to identify why the particular situation has happened. This method of data evaluation is useful to uncover data anomalies, determine the relationships within the data, and detect patterns and trends in product market analysis. Some of the diagnostic analytics used by various business firms include data discovery, alarms, drill-down, correlation, drill-up, and data mining. In-depth analysis by experienced demand planners provides assistance for better decision choices. Diagnostic analytics is a reactive process; it helps us only to anticipate the possibility of continuation of the current situation even when used with forecasting.

c) Predictive Analytics

Predictive analytics forms a part of business intelligence that uses predictive and descriptive factors of the available data to forecast and identify the possibility of the occurrence of an unknown pattern in the near future. Predictive analytics is a subset of business intelligence that analyzes and predicts the possibility of an unknown future result using descriptive and predictive factors from the past. It combines analytical techniques, data mining strategies, predictive models, and forecasting methods to assess the possibility of risk and linkages in the current data to perform future predictions. At this point, you are more interested in why something happened than in what happened. It offers proactive market responses.

d) Prescriptive Analytics

Prescriptive analytics combines descriptive, predictive, and diagnostic analysis to create the possibility to make things happen. Beginning with descriptive analysis, which informed us about what has happened, the next stage was to do a diagnostic about why it happened and the next was predictive analysis to predict when it would happen. As a consequence, prescriptive analysis uses business principles and mathematical models on the data to infer future decisions/actions from the current data. Business firms can implement prescriptive analytics in day-to-day transactions only when analytics-driven culture is followed for the entire organization. Larger firms such as Amazon and McDonald's employ prescriptive analytics to increase revenue and customer experience by increasing their demand planning.

e) Cognitive Analytics

A software that takes all data and analytics and also learns on its own without explicit human direction is cognitive analytics. To achieve this

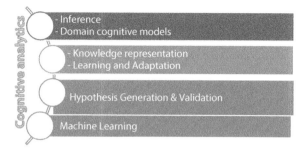

Figure 1.2 Conceptual view of cognitive computing [6].

self-learning, cognitive analytics combines advanced technologies like Natural Language Processing (NLP), artificial intelligence algorithms, machine learning and deep learning, semantics, data mining, and emotional intelligence [5]. Using these techniques, the cognitive application would become smarter and repair itself.

1.3 Conceptual View of Cognitive Systems

Internal components of the cognitive analytics engine are depicted in Figure 1.2 by the large rectangle. To represent and reason with information, many knowledge representation structures are required. A variety of machine learning methods and inference engines are also required. Domain cognitive models encapsulate domain-specific cognitive processes to facilitate cognitive style problem solving. The learning and adaptation component increases system performance by learning from prior encounters with users. In contrast to all previous analytics, cognitive analytics provides many solutions to a query and assigns a level of confidence to each response. In other words, cognitive analytics use probabilistic algorithms to provide several responses with variable degrees of relevance. Noncognitive analytics, on the other hand, uses deterministic algorithms to calculate just one solution to each inquiry. Another component, labeled Hypothesis Generation & Validation, is required to compute numerous responses [6].

1.4 Elements of Cognitive Systems

In general, the important components in Figure 1.3 of a cognitive computing system may be divided into three groups.

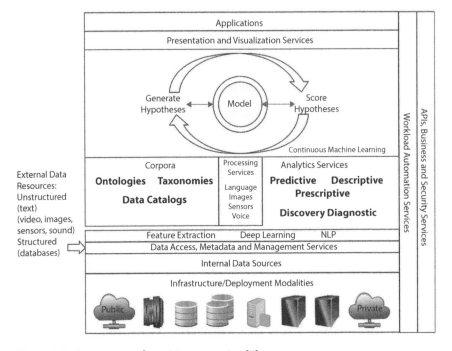

Figure 1.3 Components of cognitive computing [7].

- A method of analyzing input
- A collection of content or information supporting the conclusion
- A method of comparing the signal to the content/information corpus

Let us go a little further by looking at some of the bits that make up such components. A highly parallel and distributed infrastructure is provided along with computing and storage cloud resources.

a) Metadata, Data Access, and Management Services

It uses the data acquired from the database and its provenance along with the methods to identify the properties of those data that are not static in nature (e.g., when is the data source created, by whom, etc.). The corpus comprises internal and external data sources that prepare the data to be used within it (i.e., data have to be selected, cleaned, and accuracy monitored).

b) Taxonomies, Corpus, and Data Catalogs
Corpus contains huge amounts of data that are mainly text-based data such as documents, reports of patients, consumers, etc. It also has unstructured and semi-structured data such as videos, photos, and audio. The corpus also has information about the ontology and the connections of data. A taxonomy provides persuasion of data present in an ontology.

c) Data Analysis Services
These services are used to create knowledge about the data that have been ingested and processed within the corpus. A cognitive system model is created by a collection of sophisticated algorithms. The machine learning algorithms adopted in Cognitive Systems have two sets of dynamics, namely, (i) hypothesis generation and (ii) hypothesis evaluation. A hypothesis demonstrates observable occurrence based on the evidence and is a testable statement. The evidence to support the hypothesis is derived from a repeated process of training the data. The learning of structured and unstructured data requires suitable tools for processing it. NLP (Natural Language Processing) services may analyze and find patterns in unstructured textual input to help a cognitive system. Deep Learning (DL) technologies are required for unstructured data such as photos, videos, and audio.

d) Data Visualization
Data visualization is very much useful to interpret the results in graphical form and assist in easily drawing recommendations. The visualization of patterns and their connections with color, structure, and other forms are easier to identify and understand the pattern of data. Above all, the applications must be developed by making use of all the capabilities of the cognitive system that modulate the business in different verticals.

1.5   Features, Scope, and Characteristics of Cognitive System

Features of Cognitive Systems
The objective of cognitive computing is to design a framework that can handle complex issues without human intervention. Applying cognitive function for commercial and general application development can be done by incorporating the features proposed by Cognitive Computing Consortium. The application must have the following features:

a) Adaptive

The machine learning algorithms are used to build the cognitive system at its initial stage. The system must learn and train the system as well as learn to adapt to the surroundings as it is mimicking the human brain. A single job cannot be coded into the systems. It must be dynamic in terms of data collection, goal comprehension, and requirement fulfilment.

b) Interactive

The solution developed using cognitive systems like the human brain will interact with different kinds of systems such as processors, cloud services, mobile, and the user. The interaction with the system is bidirectional. The human input is interpreted by natural language processing and deep learning techniques to give a suitable result. This can be done by various chatbots like Mitsuku.

c) Iterative and Stateful

The cognitive system remembers the previous data input in a process and substitutes the appropriate input when the application is called in the future. This is to characterize the problem by asking questions or locating further information. This feature demands the proper use of quality data and validation procedures to substantiate that the system is sufficiently supplied with acceptable information in turn to offer reliable and up-to-date input.

d) Contextual

The cognitive system must comprehend, recognize, and draw contextual characteristics such as proper domain, user profile, synonyms, objective, time, place, regulations, process, task, and syntax. The system relies on various sources including unstructured and organized digital data, as well as sensory data (gestural, visual, aural, or sensor-provided).

Scope of Cognitive Computing

According to IBM Institute for Business Value, the scope of cognitive computing [8] involves discovery, interaction, and decision. This is much similar to the cognitive ability of human in everyday life. The simplified figure displaying the cognitive functionality is shown in Figure 1.4.

a) Engagement

Cognitive systems contain unstructured and organized data. It exerts deep domain knowledge and delivers proper expert advice. The testable statement and arguments are constructed by the model by taking into account

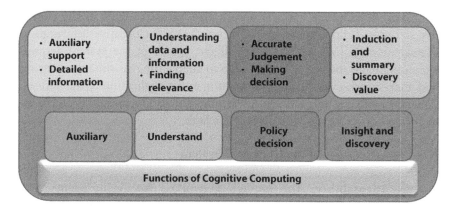

Figure 1.4 Functions of cognitive computing [9].

contextual relationships among various items in the system. This is help-ful in finding solutions to unclear and irrational facts. This leads to the capability of the system to engage people in discerning discussions such as chatbot, which is the best example of its kind. Many AI-enabled chatbots are well trained with the required domain knowledge, enabling them to be adopted in many specific business applications.

b) Decision

The cognitive systems are more ahead in taking decisions when compared to others. This can be achieved by reinforcement learning. These decisions are continuously evolving based on the latest information, recent results, and actions. The self-adaptive system has the capability to change the con-fidence score by retracing the decisions and is helpful in self-governing decisions. The best example of this kind is IBM Watson in the healthcare industry. This system acquires and analyze patient data like medical history and diagnosis. The solution given by the system must have the capacity to read the queries and, based on the complex medical data and doctor's notes, clinical comments.

c) Discovery

The next advanced level of cognitive computing is discovery. Discovery means finding insights and grasping a huge volume of information. This model is built using deep learning and unsupervised machine learning methods. The system adapts to the growth of data and should assist the people by efficiently utilizing the information. While in the early develop-ment stage of the system, certain discovery features surfaced, the benefits for future demands of application are also attractive. One of the cognitive solutions is Louisiana State University's (LSU) Cognitive Information Management (CIM) shell. The intelligent agents of the system acquire streaming data, such as text and video to enable the real-time monitoring and analysis achieved through inspection and visualization of the system. The CIM Shell not only delivers a warning, but also reconfigures on the fly to isolate a crucial event and correct the fault.

Key Characteristics of the Cognitive System

   a) **Adaptability:** The model's networked intelligent agents acquire data from various sources; the data may be in the form of documents and video, to form an interactive sensing, inspection, and visualization system that allows real-time monitoring and analysis. The CIM Shell not only gives a warning, but also reconfigures spontaneously to isolate and rectify a critical event.
   b) **Contextual Understanding:** Cognitive learning helps AI systems to perceive and interpret data in context. They can recognize important patterns, linkages, and dependencies, allowing for more accurate predictions and judgments.
   c) **Continuous Improvement:** Cognitive learning is an iterative process in which AI systems learn from input, identify areas for development, and continuously update their models and algorithms. To improve performance, they can use techniques such as reinforcement learning.

1.6   Cognitive System Design Principles

The model in a cognitive computing system emphasizes the collection of data as well as the set of techniques used to produce and score hypotheses in order to solve problems, answer questions, or uncover recent information from the corpus. What type of forecasts you can make, trends and anomalies you can spot, and actions you can take are all determined by how you model the environment. The system's designers provide the initial model, but the cognitive system updates it and uses it to give solutions to queries or give in-depth information. The corpus is the storage of knowledge used by machine learning algorithms to continually modify the model based on feedback from the user. The system performance is mainly influenced by the choice of data structure since it is repeated for the purpose of accomplishing hypotheses and retrieval of information. The design process is carried out with the common workloads before being implemented to particular architectures.

A cognitive system is intended to anticipate future outcomes by using a domain model. A cognitive system is created with a number of phases. It necessitates comprehension of the sorts of queries that must be asked, accessible data, and the development of a data source large enough to support the production of hypotheses about the domain based on observable facts. As a result, a cognitive system is designed to analyze alternate hypotheses, generate hypotheses from data, and decide the availability of supporting evidence to solve issues. A cognitive system can give end users a strong method to learn and train the system by employing machine learning algorithms, question analysis, and advanced analytics on relevant data, which may be unstructured or organized.

1.7   Backbone of Cognitive System Learning/Building Process [10]

a) Ontology vs. Taxonomy

Taxonomies give machine-ordered representations. The W3C [11] refers to ontology as a collection of terms that are more complex in nature. Taxonomy is the organization of different items or classes. It is observed that taxonomies:

- Use a hierarchical framework and give identifying name for every item with respect to other objects.
- Record membership attributes of every object with respect to other objects.
- The objects in any domain are classified or categorized based on distinct criteria. These guidelines must be consistent, comprehensive, and clear.
- Ensure that the newly added item must fit into any of the objects or categories and include rigor specifications for the new item.
- The characteristics of the class above it are derived and it may also have additional attributes.

Ontology is a subset of taxonomy; however, it contains additional information about the behavior of things and their connections. Ontologies take into account how a domain impacts components such as model selection, rules, representations, and needed operations. Modeling the style of thinking taxonomy alone is not enough. The incorporation of ontology in representation presents the information to the user in simple terms after processing it. Thus, the Ontology Web Language (OWL) is promoted to present the content to the user, in which AI may comprehend complex items, and ontology is used for this purpose. OWL adds vocabulary to formal semantics, allowing for increased machine interpretability of material. A cognitive system can generate its own internal representations and understandings of ideas in a data-driven way by using techniques such as deep learning, reinforcement learning, or other cognitive architectures, without explicitly relying on pre-defined taxonomies or ontologies.

b) Using Taxonomies and Ontologies

The usage of ontology and taxonomies enables the system to learn efficiently, and the tools used to apply these in system architecture need to be investigated. To design a system with cognitive activities then, the concept that comes to mind is taxonomies. There are a variety of data structures used to store taxonomies. The structure may be a taxonomy tool, draught database, or relational database. Taxonomies and ontologies serve as the foundation for computer self-learning, opening the door to previously inconceivable and useful cooperation with machines.

c) How Do Cognitive Agents Become "Smarter" Than Before/Learning of Cognitive Systems

The cognitive agents become smarter after repeated processing of data from which the system learns by itself through the system that can give accurate results. There are several techniques available for the learning process and they are depicted in Figure 1.5.

i.   **Supervised Learning for Beginners**
     Artificial intelligence plays a very important role in the cognitive systems learning process. In the learning phase, systems are trained using supervised machine learning algorithms to recognize the association among the available data. When presented with fresh data, the system tries to relate the new data based on the pattern rule and constantly improves its decision by developing the ability to relate new patterns by updating the additional pattern finding rules into its knowledge base. This type of artificial

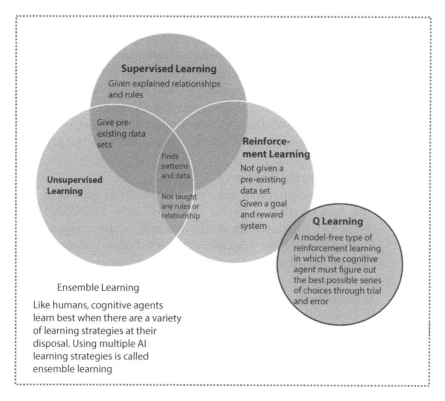

Figure 1.5 Types of learning [4].

intelligence is sometimes used in business chatbots built to answer commonly asked questions through the phone or instant messenger.

ii. **Unsupervised Learning for Beginners**

Unlike supervised learning, the cognitive system is not fed with rules based on the available data in the learning phase; instead, the system should educate itself on the association between the available data and frame rules and modify it in the course of time based on the learning experience.

iii. **Reinforcement Learning for Beginners**

Unlike supervised and unsupervised learning, the system is not provided with patterns or data. The reinforcement learning is provided with reward points and goals. The system aims to maximize its reward by learning its surroundings. Data discovery and pattern identification from the environment data in the learning process assisted through a reward-seeking approach. Consider a hide-and-seek game playing example, The cognitive system plays games against itself millions of times to learn the game. More complex approaches are learned based on the problem like erecting barriers. In another illustration of AI's divergent thinking, the cognitive agents discovered flaws inside the physical restrictions of their simulation. This enabled them to create gaming tactics that their developers had not expected.

iv. **Q-Learning for Beginners**
The system learns through a model-free learning approach. The system starts with no rules and zero knowledge about the surroundings in the learning phase; instead, all possible options available are given to the system to explore. Based on trial-and-error actions, the system has to decide the best option. It is like searching for a place that is available in a locality without GPS or a map. The system discovers the path itself through trial and error. Following that, it learns how to optimize its travel path and speed to that place by drawing on previous experiences.

v. **Combining Learning Strategies**
Each technique has its own limitations and advantages. The more human-like thinking can be made possible by ensembling the learning techniques. The learning methodologies allow the cognitive system to learn from the available data based on the application, Furthermore, we enable the AI to learn from its own experiences and build its own rules concerning the problem-specific strategies by utilizing reinforcement learning and Q-learning. When it comes to SDi's AI military simulation, we have combined the aforementioned learning methodologies into our ensemble methodology with the objective of building a cognitive agent that can one day outperform even the finest military tacticians. Supervised and unsupervised learning approaches, for example, enable us to feed the AI data sets from pre-existing military simulations with real-world battle data.

These learning methodologies also allow us to teach the cognitive agent the basic laws of human fighting and intellect, such as fundamental spatial knowledge. Furthermore, we enable the AI to learn from its own experiences playing against top-tier military specialists and build its own rules concerning ideal fighting strategies by utilizing reinforcement learning and Q-learning. Unsupervised learning cognitive agent is widely used in personalized suggestions like Netflix movie recommendations or Amazon product suggestions or personalized advertisements encountered while browsing through social media.

1.8   Cognitive Systems vs. AI

Many everyday goods and services, ranging from search-engine advertising applications to face recognition on social networking sites to "smart" automobiles, phones, and electric grids, are beginning to display characteristics of Artificial Intelligence (white paper). Cognitive systems, on the other hand, integrate five fundamental skills.

a) **They Foster Deeper Human Engagement**
   Cognitive systems provide more complete human connections with individuals depending on the mode, form, and quality that each person chooses. They sort through all of this organized and unstructured data to determine what is truly important in engaging a person. These encounters grow more natural, anticipatory, and emotionally appropriate as they continue to learn.

b) **They Scale and Elevate Expertise**
   Knowledge in every business and profession is growing faster than any practitioner can keep up with—journals, new procedures, new regulations, new practices, and totally new areas. A notable example may be seen in healthcare, where it is estimated that it took 50 years in 1950 to double the world's medical knowledge; seven years in 1980; and less than three years in 2015. Meanwhile, in his or her lifetime, each individual will create one million terabytes of health-related data, the equivalent of almost 300 million books. Cognitive systems are intended to assist organizations keep up with the times by acting as a companion for professionals looking to improve their performance.

c) **They Infuse Products and Services with Cognition**
   Cognition allows new types of goods and services to perceive, reason, and learn about consumers and the environment around them. This enables for ongoing refinement

and adaptability, as well as augmentation of their capabilities to provide previously unimagined uses. This is already occurring with vehicles, medical gadgets, appliances, and even toys. The Internet of Things is drastically increasing the range of digital products and services—and cognition can now move where code and data go.

d) **They Enable Cognitive Processes and Operations**

Cognition also changes the way a business functions. Business operations with cognitive capabilities capitalize on the phenomenon of data from both internal and external sources. This increases their knowledge of processes, context, and surroundings, leading to continuous learning, improved forecasting, and enhanced operational performance—as well as decision-making at the pace of today's data.

e) **They Enhance Exploration and Discovery**

Finally, the most effective weapon that cognitive firms will have is considerably better "headlights" into an ever-turbulent and complicated future. Such flashlights are growing increasingly crucial as executives in many industries are forced to make large bets—on medication discovery, complicated financial modeling, materials science breakthrough, or founding a business. Leaders may reveal patterns, possibilities, and actionable ideas by applying cognitive technology to massive volumes of data, which would be practically hard to discover with conventional research or programmed systems alone. Cognitive computing is the third computer age that uses deep learning algorithms and large data analytics to address very important problems.

1.9   Use Cases

Cognitive computing is based on the scientific principles of artificial intelligence and signal processing. It uses machine learning, reasoning, natural language processing, speech recognition and vision, human–computer interaction, dialogue, and narrative generation, and is integrated into many cross-discipline platforms to improve decision-making. It is modeled based on how the human brain senses, reasons, and responds to stimuli. It is capable of customizing its functionality based on individuals and varying environments. Cognitive computing applications are more effective and influenced by their design. One temptation, however, is to pursue cognitive technology for the technology's sake. AI technology can assist in searching for a variety of meanings hidden deep in data, but cognitive computing will assist in making sensible judgments. Thus, cognitive computing is the way to go to make complex judgments. The success and failure of the applications depend on how it is started. Most losses are seen when starting with the technology instead of the business case. People get excited when they can do so many things with cognitive technology. Nevertheless, focusing on what impacts the bottom line would be best. Cognitive computing system applications are numerous and varied.

In 2011, manufacturing floors throughout the country welcomed a new employee: Baxter, a six-foot-tall, 300-pound robot with two long, dexterous arms and a set of expressive digital eyes that followed its arms everywhere they went. Unlike previous industrial robots, Baxter was collaborative because of cognitive computing—an AI method aimed to imitate the human reasoning process. That is how it was taught. Humans could hold its arms and show it how to perform jobs more effectively, acting as a mentor to Baxter, who could then perfect those skills. Unfortunately, Baxter's life was cut short. After much early hype, its inventor, Rethink Robotics, struggled for years to enhance its operations. IBM's Watson as in Figure 1.6, a cognitive computing system that conquered Brad Rutter and Ken Jennings on Jeopardy!, was the first cognitive computing demonstration published in February 2011 to the world; it was the cause of the end of the so-called artificial intelligence winter [4]. Watson assists doctors with diagnosing challenging patients in the world of medicine. Peer-reviewed research has been integrated into the system along with medical textbooks and journal articles. When making his most likely diagnosis, Watson refers to these sources. Natural language processing (NLP) is a tool that Watson utilizes to assist the banking sector with finding pertinent information and insights more quickly, facilitating decision-making, and enhancing client experiences. It improves operational effectiveness, analysis, and credit initiation through enterprise-wide data management. IBM's Watson with its natural language processing (NLP) and AI capabilities is perhaps the most well-known "face" of psychological registration; the term subjective processing is coined to describe frameworks that can learn, reason, and communicate in a human-like manner.

Cognitive computing holds a data analysis behavior that increases cognitive capacity by continually altering and gaining information from the data in an interactive manner [12]. The study builds a cognitive model that integrates cognitive computing and deep belief network algorithms. The developed cognitive model acts as a control system for collaborative robots.

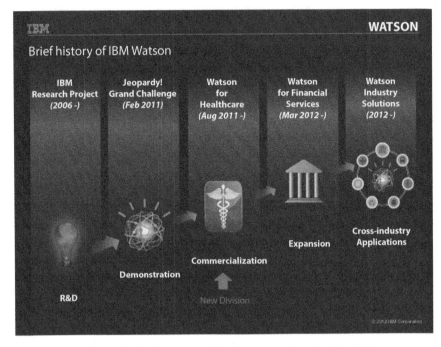

Figure 1.6 History of IBM Watson (source https://andrewlenhardtmd.com/blog/wp-content/uploads/2017/07/IBM-Watson-1.jpg).

A collaborative robot control system's performance has improved by combining cognitive computing technology and a deep belief network algorithm [13]. The edge computing vision and the cognitive computing success laid the foundation for a promising technology approach named edge cognitive computing (ECC). Cognitive computing is implemented at the network edge. Edge cognitive computing saves computing resources and provides energy-efficient service to the user with ultra-low latency [14]. The observations established that the ECC realizes an optimized human-centered behavior prediction to assist in service migration considering the network traffic and environment network resource availability.

Cognitive computing applications in smart cities have seen prominent growth and received much attention. The three defining properties of big data, namely, velocity, variety, and volume, have been handled using data analytics, machine learning, and recommender systems. It also explored how the cognitive applications side of smart cities was driven by deep reinforcement learning and its semi-supervision. The data analytics and machine learning algorithms' role in different aspects of the smart-city application projects framework, like smart environment, parking, and transportation, is analyzed to discourse the challenges in big data [15].

Life sciences have always been demanding and need rapid advancements to improve people's well-being. Traditional medicines offer a steady foundation for treating individuals, but they cannot fulfill the requirements of today's unhealthy lifestyle. Even though new technologies are penetrating themselves across healthcare communities, cognitive computing has also made its way into the field to address the constraints of traditional medicine [16]. With integrity and intelligence, cognitive systems may be implemented to solve a wide range of applications in healthcare systems. Figure 1.7 depicts the life cycle of cognitive computing systems in cyberspace [17]. The life cycle consists of different stages like data collection and transformation followed by data analysis and response. The response received is added to cyberspace as a fresh interpretation.

Figure 1.8 depicts a cognitive computing system architecture [18]. The architecture defines the integration of the public cloud, database tools, and tensor flow for the development of cognitive applications. Cognitive data transmission technique (CDTM) in health care uses simulated annealing to perform cognitive tasks and uses map reduction to perform data analytics. The cognitive EEG approach to detect pathology retains the two convolution neural network approaches used by the EEG model for pathology classification, including AlexNet and VGG 16. The CDTM collects the EEG signals and other medical data using appropriate sensors and transfers

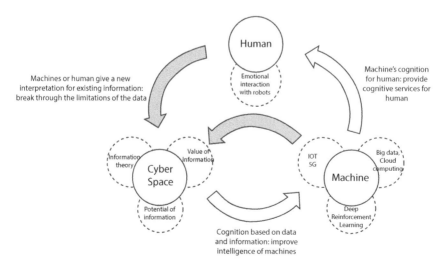

Figure 1.7 Human-centered cognitive cycle [17].

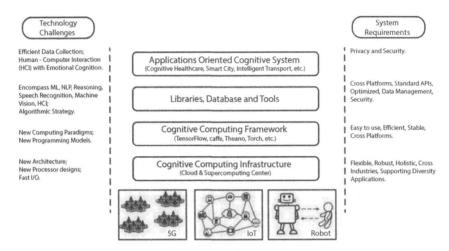

Figure 1.8 Cognitive computing system architecture [18].

them to the cloud infrastructure. The data sent to the cloud is processed using a deep learning model in cognitive systems to detect pathology. The report generated by the cognitive model is populated with the required medical practitioners for further treatment of patients on time [19].

Personalized medicine may create individualized pharmaceuticals that assure exact drug combinations and consistent dose recommendations. Personalized medicine is a scientifically proven concept in the medical domain. It aids in the early diagnosis of potential health difficulties, giving medical professionals adequate time to implement preventative measures and save patients' lives [20]. A Personal Medical Digital Assistant (PMDA) performs cognitive screening and assists healthcare professionals in trauma centers. The PMDA model is dynamic and flexible to perform situation-aware trauma tracking. This model interacts with hospital cognitive service to adjust itself to the patient's current health situation [21]. Healthcare collaboration tools are inherently complicated. As a result, it takes a long time to train individuals how to utilize these cutting-edge instruments. Cognitive computing requires a massive amount of data related to each patient to record accurate findings about a particular patient [22]. The cognitive model algorithms fail in sensor-based machines when there is a lack of a sufficient amount of data. Preserving patient privacy on data in the healthcare industry is a trivial challenge in applying cognitive computing. As the future landscape for research avenues in health care, the cognitive model ECC can be converted to a 5G cognitive healthcare model, including emotion detection and intelligent sensors to collect body signals to provide adequate healthcare [23].

In the current era of the digital age, the usage of wireless network devices is witnessing exponential growth [23]. Intelligent networking systems are prone to cyber-attacks, a significant challenge in today's environment. Cybersecurity is a barrier that protects the cyber system from cyberattacks [24]. The dynamic nature of the network demands a self-contained system capable of making immediate, precise decisions than relying on solutions based on predefined static scenarios [25]. Security still needs to be improved despite technological advancement in networking and various protection methodologies adopted in smart devices. There is continuous research for quality protocols and standards that will safeguard the data generated and stored with efficient sharing [26]. Foreseen solutions to the challenges in cybersecurity can be addressed using cognitive technologies. It plays a prominent role in identifying the behavioral pattern of cybersecurity attacks by understanding human psychological behavior. Linguistic biometric threshold schemes facilitate data exchange that makes use of cognitive processes. This method's key benefit lies in its data handling and flexible architecture. The hierarchical, layered, and mixed architecture, along with multiple levels of data handling at cloud, fog, and basics, are the main advantages of this scheme [27].

A cognitive dynamic system (CDS) is used in smart grid networks to detect false data injection (FDI). FDI is a serious issue to be considered as it may lead to blackouts and dangerous consequences in the smart grid electrical network. IEEE 4 bus and 14 bus distribution networks are used to carry out the simulation. Bayesian approach-based reinforcement learning is used to perform data analysis on previous and current perception–action cycles. The system's stability without degradation is maintained using Bayesian filtering even when multiple operations are carried out [28]. Bayesian filtering-based predictive analysis to develop cognitive dynamic systems hold human-like behavioral skills in handling direct grid current estimation models and grid networks [29]. Most of the approaches need more actual deployment in real time and exist in the literature as a theoretical framework [30].

Cognitive IoT architecture for smart buildings shown in Figure 1.9 uses several sensors. The integrated data obtained from different sensors are analyzed to avoid catastrophic accidents. The CIoT architecture consists of three essential components, edge, platform, and enterprise, combined with cognitive computing techniques [31]. The topmost layer in CIoT is the cloud infrastructure accessed through the platform as a service (Paas). The middle platform layer provides API support to facilitate user interaction for machine learning. The web interface is used to populate the data among the layers. Different application accesses are made more accessible

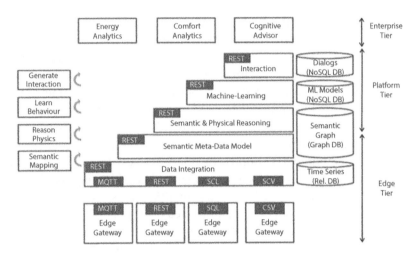

Figure 1.9 High-level cognitive IoT architecture [31].

through PaaS. It comprises three fundamental tiers: enterprise, platform, and edge, deployed with cognitive computing techniques [31]. For reasoning, the platform tier enables APIs, user interactions, and machine learning. Every layer is deployed at the cloud/edge level. The cloud's top layer provides a platform as a service (PaaS). With the web service interface, data are exchanged among various groups. PaaS facilitates client access to many apps.

People-centric cognitive Internet of things (PIoT) holds two layers, namely, the local resource layer and the cloud layer, with three essential nodes: cloud coordinator, data source, and people node. PIoT provides a device-to-device interface managing CPS service and human data. The cloud layer analyzes data using an ML algorithm. Using cognitive computing, PIoT can also provide a report on the particulate matter in the air and the degree of exposure of each user to it [32].

Cognitive computing, a new AI idea that copycats the human brain's thinking process, is gradually thriving in the automation of Industry 4.0. The problem of "data island" can be solved by federated learning with efficient processing and privacy-preserving. The blockchain-based federated knowledge obtains faster convergence through a completely decentralized approach, making the learning model self-resilient to poison attack. A decentralized cognitive computing paradigm is established in the Industry 4.0 model by combining federated learning with blockchain. The performance improvement in federated learning with the inclusion of CC for

Industry 4.0 increases its accuracy, resistance to poisoning attacks, and incentive mechanism [33].

Blockchain-based architecture is proposed to offer private and secure spatiotemporal innovative contract services for the long-term Internet of Things (IoT)-enabled sharing economy in mega smart cities. The framework provides a long-term incentive mechanism that might support safe, smart city services such as smart contracts, sharing economy, and cyber-physical communication with IoT and blockchain [34].

A Recommender System with cognitive intelligence should be capable of learning from experienced domain experts in the context of recommendation. Recommender Systems have traditionally been thought of as e-commerce product recommenders (e.g., Amazon and eBay), playlist generators for video/music services (e.g., Netflix and Spotify), or social content recommenders (e.g., Facebook and Twitter). In modern organizations, the recommender system, on the other hand, is heavily data/knowledge-driven and depends on cognitive elements such as the attitude of users, behavior, and personality [35].

a) How Can a Cognitive Application Change a Market?
Designing an application is tough when industries are changing and face new competitive challenges. Traditional software is designed to automate procedures and handle data. When a company attempts to revolutionize a conventional sector, such as tourism or customer service, it requires sophisticated technology that enables executives to uncover new strategies and expertise. A travel firm that can determine what its consumers want will stand out. What if a travel business can predict a consumer's purchase even while the buyer is unsure? What if a customer support agent can indicate that a client's issue is caused by a partner's product within minutes rather than hours? Overthinking leads to inactivity. Many companies are over-analyzing what they should be doing using cognitive technologies. They expect a comprehensive design with assured output quality, which is different. It is preferable to begin small with a good idea and then scale out and scale up to enjoy the benefit of "cognitive technology", the word that is going to rule the future.

1.10   Conclusion

Developing a cognitive system is a multistep process that begins with a business need and progresses towards its goal with a basic understanding of the domain, objective, and data source. The backbone of a cognitive system's intelligent decision process is its data corpus, its representation, and its learning process. Various learning practices and their objectives are discussed at different times. This chapter covers a brief overview of cognitive computing, highlighting its elements, characteristics, and design principles. The coexistence of technologies in the cognitive environment and their influence on markets are discussed. We also discussed the natural capability to handle the corpus of information and give emotionally appropriate responses by analyzing the real-time data rather than using conventional artificial intelligence methods. Finally, some of the use cases are presented to give insights into the applications of cognitive systems.

References

1. Amin, S.U., Hossain, M.S., Muhammad, G., Alhussein, M., Rahman, M.A., Cognitive smart healthcare for pathology detection and monitoring. *IEEE Access*, 7, 10745–10753, 2019. Horatiu, R., Dan, A., Lidia-Cristina, B., George, B.N., Cooperative cheap automated guided vehicles, in: *2019 20th International Carpathian Control Conference (ICCC)*, IEEE, pp. 1–6, 2019, May.
2. Murata, S., Li, Y., Arie, H., Ogata, T., Sugano, S., Learning to achieve different levels of adaptability for human–robot collaboration utilizing a neuro-dynamical system. *IEEE Trans. Cognit. Dev. Syst.*, 10, 3, 712–725, 2018.
3. Chen, M., Hao, Y., Gharavi, H., Leung, V.C., Cognitive information measurements: A new perspective. *Inf. Sci.*, 505, 487–497, 2019.
4. Kelly, J.E., Computing, cognition and the future of knowing. *IBM Res.*, Oct, 9, 13, 2015.
5. Pradhan, N., Singh, A.S., Singh, A., Cognitive computing: architecture, technologies and intelligent applications. *Machine Learning and Cognitive Computing for Mobile Communications and Wireless Networks*, 25–50, 2020.
6. Gudivada, V.N., Irfan, M.T., Fathi, E., Rao, D.L., Cognitive analytics: Going beyond big data analytics and machine learning, in: *Handbook of Statistics*, C.R. Rao (ed.), vol. 35, pp. 169–205, Elsevier, 2016.
7. Hurwitz, J., Kaufman, M., Bowles, A., Nugent, A., Kobielus, J.G., Kowolenko, M.D., *Cognitive Computing and Big Data Analytics*, vol. 288, John Wiley & Sons, Inc., Indianapolis, Indiana, 2015.
8. https://marutitech.com/cognitive-computing-features-scope-limitations/
9. Lv, Z. and Qiao, L., Deep belief network and linear perceptron based cognitive computing for collaborative robots. *Appl. Soft Comput.*, 92, 106300, 2020.
10. Sanz Bravo, R., Bermejo Alonso, J., Morago, J., Hernández, C., Ontologies as backbone of cognitive systems engineering, *Annual Convention of the Society*

for the Study of Artificial Intelligence and the Simulation of Behavior (AISB), pp. 218–223, 2017.

11. https://sdi.ai/blog/cognitive-agents-reinforcement-learning-for-beginners/

12. Di Nuovo, A., Acampora, G., Schlesinger, M., Guest editorial cognitive agents and robots for human-centered systems. *IEEE Trans. Cognit. Dev. Syst.*, 9, 1, 1–4, 2017.

13. Lv, Z. and Qiao, L., Deep belief network and linear perceptron based cognitive computing for collaborative robots. *Appl. Soft Comput.*, 92, 106300, 2020.

14. Chen, M., Li, W., Fortino, G., Hao, Y., Hu, L., Humar, I., A dynamic service migration mechanism in edge cognitive computing. *ACM Trans. Internet Technol. (TOIT)*, 19, 2, 1–15, 2019.

15. Habibzadeh, H., Boggio-Dandry, A., Qin, Z., Soyata, T., Kantarci, B., Mouftah, H.T., Soft sensing in smart cities: Handling 3Vs using recommender systems, machine intelligence, and data analytics. *IEEE Commun. Mag.*, 56, 2, 78–86, 2018.

16. Saji, M., Sridhar, M., Rajasekaran, A., Kumar, R.A., Suyampulingam, A., Krishna Prakash, N., Iot-based intelligent healthcare module, in: *Advances in Smart System Technologies: Select Proceedings of ICFSST 2019*, Springer, Singapore, pp. 765–774, 2021.

17. Pradhan, N., Singh, A.S., Singh, A., Cognitive computing: architecture, technologies and intelligent applications. *Mach. Learn. Cogn. Computing Mobile Commun. Wireless Networks*, 25–50, 2020.

18. Gunasekhar, T. and Teja, M.S., *Cognitive engineering for next generation computing: A practical analytical approach*, IEEE, pp. 189–217, 2021.

19. Amin, S.U., Hossain, M.S., Muhammad, G., Alhussein, M., Rahman, M.A., Cognitive smart healthcare for pathology detection and monitoring. *IEEE Access*, 7, 10745–10753, 2019.

20. Cirillo, D. and Valencia, A., Big data analytics for personalized medicine. *Curr. Opin. Biotechnol.*, 58, 161–167, 2019.

21. Montagna, S., Mariani, S., Gamberini, E., Ricci, A., Zambonelli, F., Complementing agents with cognitive services: A case study in healthcare. *J. Med. Syst.*, 44, 1–10, 2020.

22. Ravindran, N.J. and Gopalakrishnan, P., Predictive analysis for healthcare sector using big data technology, in: *2018 Second International Conference on Green Computing and Internet of Things (ICGCIoT)*, IEEE, pp. 326–331, 2018, August.

23. Sreedevi, A.G., Harshitha, T.N., Sugumaran, V., Shankar, P., Application of cognitive computing in healthcare, cybersecurity, big data and IoT: A literature review. *Inf. Process. Manage.*, 59, 2, 102888, 2022.

24. Parthasarathy, R.R., Wyant, D.K., Bingi, P., Knight, J.R., Rangarajan, A., DeTER Framework: A novel paradigm for addressing cybersecurity concerns in mobile healthcare. *Int. J. Intell. Inf. Technol. (IJIIT)*, 17, 2, 1–24, 2021.

25. Wyant, D.K., Bingi, P., Knight, J.R., Rangarajan, A., Deter framework: A novel paradigm for addressing cybersecurity concerns in mobile healthcare. *Int. J. Intell. Inf. Technol. (IJIIT)*, 17, 2, 1–24, 2021.
26. Rani, R. and Singh, H., Fingerprint presentation attack detection using transfer learning approach. *Int. J. Intell. Inf. Technol. (IJIIT)*, 17, 1, 53–67, 2021.
27. Ogiela, L. and Ogiela, M.R., Cognitive security paradigm for cloud computing applications. *Concurrency Comput.: Pract. Exper.*, 32, 8, 5316, 2020.
28. Oozeer, M., II and Haykin, S., Cognitive dynamic system for control and cyber-attack detection in smart grid. *IEEE Access*, 7, 78320–78335, 2019.
29. Haykin, S., Fuster, J.M., Findlay, D., Feng, S., Cognitive risk control for physical systems. *IEEE Access*, 5, 14664–14679, 2017.
30. Garrett, M.A., SETI reloaded: Next generation radio telescopes, transients and cognitive computing. *Acta Astronaut.*, 113, 8–12, 2015.
31. Ploennigs, J., Ba, A., Barry, M., Materializing the promises of cognitive IoT: How cognitive buildings are shaping the way. *IEEE Internet Things J.*, 5, 4, 2367–2374, 2017.
32. Yang, L., Li, W., Ghandehari, M., Fortino, G., People-centric cognitive internet of things for the quantitative analysis of environmental exposure. *IEEE Internet Things J.*, 5, 4, 2353–2366, 2017.
33. Qu, Y., Pokhrel, S.R., Garg, S., Gao, L., Xiang, Y., A blockchained federated learning framework for cognitive computing in industry 4.0 networks. *IEEE Trans. Ind. Inf.*, 17, 4, 2964–2973, 2020.
34. Rahman, M.A., Rashid, M.M., Hossain, M.S., Hassanain, E., Alhamid, M.F., Guizani, M., Blockchain and IoT-based cognitive edge framework for sharing economy services in a smart city. *IEEE Access*, 7, 18611–18621, 2019.
35. Beheshti, A., Yakhchi, S., Mousaeirad, S., Ghafari, S.M., Goluguri, S.R., Edrisi, M.A., Towards cognitive recommender systems. *Algorithms*, 13, 8, 176, 2020.

Cognitive Computing System-Based Dynamic Decision Control for Smart City Using Reinforcement Learning Model

Sasikumar A.[1], Logesh Ravi[2], Malathi Devarajan[3], Hossam Kotb[4] and Subramaniyaswamy V.[5*]

[1]*Department of Data Science and Business Systems, Faculty of Engineering and Technology, SRM Institute of Science and Technology, Kattankulathur, Tamil Nadu, India*
[2]*Centre for Advanced Data Science, Vellore Institute of Technology, Chennai, India*
[3]*School of Computer Science and Engineering, Vellore Institute of Technology, Chennai, India*
[4]*Department of Electrical Power and Machines, Faculty of Engineering, Alexandria University, Alexandria, Egypt*
[5]*School of Computing, SASTRA Deemed University, Thanjavur, India*

Abstract

Artificial intelligence (AI) is currently implemented to support many human-centered solutions, including medical, automated mobility, and other areas. As an emerging application developed with the help of AI, cognitive computing offers individualized connections and features that mimic human behavior for interpersonal interactions. On the other hand, much data are produced by applications of smart cities like healthcare, intelligent transportation, the retail sector, and disaster response. Properly managing the massive amount of created data is a constant concern. The analysis of vast amounts of data using cognitive computing has recently been the topic of many existing studies. These studies still need to address key issues, especially the portability and adaptability of data collected in a smart city context. Data gathered through thousands of sensors can be integrated into various cognitive computing systems to ensure real-time response. In this chapter, we presented the cognitive computing layer and suggested architecture for smart city applications. The proposed architecture was developed to analyze

Corresponding author: vsubramaniyaswamy@gmail.com

Elakkiya, R. and Subramaniyaswamy V. (eds.) Cognitive Analytics and Reinforcement Learning: Theories, Techniques and Applications, (29–50) © 2024 Scrivener Publishing LLC

data obtained from smart city applications using a reinforcement learning model. The experimental results showed that the proposed reinforcement learning-based cognitive computing has real-time decision controls in smart industry ecosystems.

Keywords: Artificial intelligence, cognitive computing, internet of things, reinforcement learning, smart city

2.1 Introduction

With the rapid growth of computing software and hardware advancements, big data, and machine intelligence, cognitive computing has recently attracted substantial attention in academia and industry. In the academic setting, cognitive computing is defined by the IEEE Technical Activity as a multidisciplinary study and application domain that draws on techniques from several academic disciplines to build machines with thinking capabilities comparable to those of the human brain. Industry-wide, the IBM Company created the cognitive system known as Watson, which could analyze and deliberate about natural language and learn from texts without guidance. Strong AI is at the center of those efforts, and the cleverness of these computers is based on the variety of data that the internet provides [1, 2].

The quantity of unstructured data is huge; we need a different kind of innovation to handle it. Researchers think that to handle this enormous amount of unstructured data, designers now need a new type of system. In the era of cognitive computing and interacting with everything without cognitive decision power, intelligent systems are worthless. Designers must work with cognitive science and the intelligent system to make the IoT more intelligent to address all of the modern age's prospective difficulties and meet all of its prerequisites [3]. "Cognitive Internet of Things" refers to this fusion of cognitive science with IoT. Let us use a real-world example to illustrate the CIoT ideas [4]. Imagine drivers driving through a city for the first time and knowing nothing about it. In the present day, traffic congestion in a modern metropolis bothers everyone. When drivers tell their cars where they are going, an intelligent CIoT will automatically provide you with the route. Not only the direction but also the shortest and least-trafficked route should be considered. The vehicle will contact the main data center, where the entire city's information is accessible. The cloud infrastructure will evaluate the most recent city statistics and provide the best route to the driver's destination.

The increasing rise in data science and intelligent systems gives some innovative solutions to smart city applications [5]. Deep learning (DL), which trains cognitive behavior from available device data, has particularly benefited smart cities [6]. Interpretation, understanding, learning, decision, rational thought, preparing, layout, and pixel density are all parts of DL. The network can portray, gain knowledge, and argue thanks to the smart city and DL. Individuals can gain knowledge from new data processing easily, but it is challenging for computers to rapidly modify their understanding if the input information changes suddenly and unexpectedly.

Cognitive technology aims to replicate and digitize human learning processes [7]. The DL employed by edge devices like the Nvidia Jetson TX aims to mimic human brain functions. Unstructured data can be retrieved from its whole peripheries using cognitive technology. A semantics library comprising ideas, objects, and linkages was constructed to provide a real dataset with the industry system. Cognitive technology allows the machine to view the world abstractly. This is helpful when evaluating a machine-learning model's performance. The smart city cognitive data streams give network services the ability to respond to outside occurrences adaptively, exactly like a human would. Hence, intelligent applications that go beyond interactions at the human level are also feasible.

This work aims to tackle the problem of the adaptability and sustainability of data in developing cognitive computing-based technologies in a smart city context. Many organizations widely recognize the possibility of offerings that go further than traditional AI-based remedies, and cognitive computing represents an upcoming evolution that is expanding quickly in the industry [8]. Large amounts of streaming data are generated by smart cities, which aids in improving the effectiveness and efficiency of current cognitive computing-based solutions. A design for a smart city setting must permit numerous and shared streams of cooperative contextual data about people and the environment to support numerous cognitive computing-based implementations. The key contribution and relevance of this study are given below:

- We present the foundational innovations for cognitive computing, such as the reinforcement learning model. The device has identified the worth of established cyberspace data by implementing this model. We also introduce the idea that, for cognitive computing to succeed in improving machine intelligence, it will need to use both traditional methods of analyzing data and novel approaches to overcoming data constraints.

- We introduce a dynamic control model based on a reinforcement learning model for real-time data processing applications. We predict that combining data and the interaction between humans, machines, and cyberspace will allow machines to render more intelligent services to humans due to these amalgamations.

The rest of this chapter is divided into the following sections. The smart city component is described in Section 2.2 Section 2.3 describes the related work of cognitive computing, Section 2.4 introduces the fundamental technologies of cognitive computing, and Section 2.5 illustrates applications based on smart city cognitive computing. Lastly, Section 2.6 provides the conclusion of the chapter.

2.2   Smart City Applications

A smart city is made up of numerous elements, which are represented in Figure 2.1. Implementations for smart cities typically have four components: data gathering, transfer and reception, preservation, and evaluation. Data gathering is the first component. Application-specific data collecting has been a major force behind sensor development across various industries [9]. Data transfer from the data-gathering devices to the cloud for storage and analysis is the second component of the process. This goal has been accomplished in several ways, including using 4G and 5G technologies, city-wide Wi-Fi networks, and numerous local network types that can transmit data locally and globally [10].

The third phase is storage space, where various storage strategies are employed to organize and process data in the fourth stage, analysis techniques [11, 12]. Data analysis involves extracting correlations and interpretations from the obtained data to help decision-making. In some circumstances, straightforward assessments, such as simple decisions and aggregation, would also be effective. The accessibility of the cloud enables real-time analysis using statistical techniques, machine learning, and deep learning algorithms, as well as diverse data collection, preservation, and computation for more complicated judgments [13].

a) Smart Farming
Food production is one of the most crucial elements of the 2030 Sustainability Building Targets of the United Nations. The race to guarantee that the food industry is made feasible and that shrinking resources

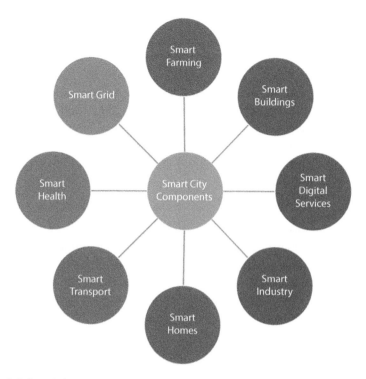

Figure 2.1 Smart city components.

like water are utilized efficiently has been a primary concern for countries worldwide due to the growing global population and rapidly deteriorating climate change causing unpredictable weather in food-producing regions. Using sensors in plants and fields to assess different metrics to aid decision-making and avoid diseases, pests, and other issues is known as smart agriculture [14]. An aspect of the smart agriculture concept is precision agriculture, which includes sensors installed in plants to generate specific observations, enabling tailored care procedures to be applied. Sustainable food production will depend on precision farming; hence, it is crucial to the struggle for environmentally friendly agriculture [15]. Crop surveillance and detection of diseases, as well as information crop care and judgments are the two primary uses of intelligence in IoT for agriculture.

b) Smart Digital Services
The activities that keep a city's people alive include municipal responsibilities like water delivery, waste treatment, climate control and monitoring, etc. Smart city services also include transportation services. Water quality

sensors can be installed to continuously provide information about the standard of water utilized in the city and to find leaks [16]. The management of waste is a common aspect of smart city initiatives. It has been incorporated into many of the initiatives previously mentioned, from chutes in Barcelona to bins outfitted with sensors and connected to the cloud to alert the appropriate authorities when they need to be emptied and use AI to determine the best route to minimize costs. Moreover, sensors can direct city dwellers to the next available free parking place to reduce gasoline expenses and monitor environmental conditions in a city to measure pollution levels [17].

c) Smart Grid
The main generating source for most electrical lines, often a hydropower or fossil energy power station, only provides energy in one direction. The energy production scheme used with these systems requires that the energy generated by these source materials surpass the requirement by a significant amount to ensure an uninterrupted power supply because electricity production is controlled via responses from the transformer, but there is no knowledge input from the customer side. Determining flaws in these systems and fixing them is likewise a time-consuming task. In addition, as renewable energy technologies become more affordable, consumers now produce their energy and receive supplies from the major utility. The term "smart grids" refers to using ICT technologies to increase the observability of existing and newly installed grids, enable decentralized electricity production at both the consumer and utility ends, and add self-healing abilities to the grid [18]. At various grid locations throughout the supply lines up to the consumer, real-time power data are provided to utilities. Smart grids enable the integration of various energy sources and the self-healing [19] of the network to ensure an uninterrupted supply and better control of power production using predictive models generated from gathered consumer data.

d) Smart Health
The term "Smart Health" describes the implementation of IT to enhance the accessibility and quality of medical services. This topic has received much attention from scholars and healthcare practitioners due to rising healthcare expenses and a growing global population. Existing health systems are overloaded and so cannot accommodate the increasing demand from the public. With the increasing use of cell phones and health monitors [20], they can collect basic information on people's health, capture everyday activities, and identify abnormal movements using vestibular sensor technology.

Smart health intends to ensure that treatment is accessible to as many individuals as possible via telehealth services and enhanced diagnostics aid to doctors using machine learning and artificial intelligence [21]. Hence, minimize the overall costs as well as stress on healthcare systems.

e) Smart Home

A smart home employs cutting-edge technology to automate and optimize numerous operations and processes within the home. This can involve using a central hub or smartphone app to control lighting, heating, and cooling systems; security cameras; door locks; appliances; and entertainment systems. Smart home technology can give several benefits, such as enhanced energy efficiency, convenience, and improved home security. A smart thermostat, for instance, can learn your preferences for heating and cooling and adjust accordingly, saving you money on your energy costs. Smart door locks let you grant entry to your house from a distance, and security cameras can notify you via text message if any suspicious behavior is spotted [22]. Several platforms, such as Apple HomeKit, Google Assistant, and Amazon Alexa, may manage many different kinds of smart home gadgets on the market. Smart home gadgets are becoming more accessible and affordable as technology develops, making it simpler for homeowners to incorporate them into their daily lives.

f) Smart Industry

Industries all over the world are actively working to become more productive and efficient while lowering costs. The concept of an interconnected factory with all of its intermediary functionaries fully integrated and cooperating with one another is at the heart of the Industry 4.0 model. The internet of things has made this feasible [23]. The industry has benefited from the use of IoT in factory and fabrication processes, as well as cyber-physical systems incorporating employees and machineries, including quicker and more effective innovation, optimization of manufacturing schemes (resources and processes), better product quality, and increased safety for factory workers [24]. Nevertheless, interacting with a variety of heterogeneous machines and devices poses its own set of difficulties, necessitating cyber-physical systems with design adaptability, accessibility, and rapid installation for use in IoT applications for smart industries [25]. IoT and artificial intelligence have worked together to accelerate the creation and implementation of Industry 4.0 services. Data from these resources provide a possibility for employing AI approaches to boost automation, carry out business intelligence activities, and more because sensors are integrated in machines and other manufacturing processes. In fact, for a smart sector, researchers have proposed structures for successfully integrating inside of IoT. Proactive maintenance, surveillance diagnosis, and production planning are where artificial intelligence shines most in the manufacturing sector.

g) Smart Buildings

The term "smart infrastructure" describes a network of electronic and mechanical parts that work together to track and improve the condition of essential infrastructure, including roads, bridges, and power plants. The smart infrastructure uses sensors, data analytics, and machine learning algorithms to increase efficiency, lower costs, and increase sustainability. Smart buildings use sensors to regulate heating and cooling depending on occupancy and weather patterns; autonomous vehicles optimize traffic flow and reduce traffic; and smart energy networks incorporate renewable energy sources and balance supply and demand. Technology advancements like the Internet of Things (IoT), artificial intelligence (AI), and cloud computing are driving the development of smart infrastructure [26]. It has the potential to revolutionize how we plan, construct, and maintain our infrastructure, improving our cities' livability, sustainability, and resilience.

h) Smart Transport

Traffic, noise, timing, and cost-reduction challenges for public transportation are among the traffic-related issues that plague many metropolitan areas. Communication between vehicles, infrastructure, and pedestrians is becoming routine thanks to the rapid development and adoption of new information and communication technologies. These technologies, including vehicular communication systems, overhead RFID readers and cameras, mobility devices, and pedestrian infrastructure, have made it possible to create intelligent transportation systems. Many methods employ GPS data to track driver behavior and traffic patterns since cars have GPS devices and every motorist owns a mobile [27]. These real-time data are already used for trip planning in public transportation and route mapping in apps like Waze and Google Maps. Drivers might also be directed to the closest available spot of free parking via parking systems with sensors.

2.3 Related Work

In this section, we briefly review cognitive computing and various IoT architecture related to smart cities. Also, we discussed the reinforcement learning model for cognitive systems and IoT architecture for smart cities as they pertain to the creation of interoperable smart systems.

a) Cognitive Computing for Smart City

The term "cognitive smart city" describes the integration of new IoT and smart city technologies and the big data and AI methods generated by these developments. IBM Watson, one of the industrial products moving towards cognitive structures, provides a cognitive system with several statistics and machine learning solutions based on continuous learning (i.e., the learning process is improved in future rounds depending on the feedback from past rounds). IBM defines cognitive computing as the ability of computers to learn from a variety of datasets, deliver explanations, communicate with humans in normal languages, and pick up knowledge from the context in which they are utilized.

Several published works in the academic literature recommend cognitive solutions tailored to the requirements of IoT-based frameworks. In the setting of smart cities, Vlacheas *et al.* [28] suggested a cognitive management structure to allow intelligent items to link to the most relevant things, thus increasing the value to the end user. Virtual items, composite virtual items, and the service level are the three tiers of their structure, all devoted to reusing existing objects' functionality and services. From the service level, stakeholders and applications can infer the features that the service must support. CVOs are tasked with carrying out these duties on behalf of their organizations. The authors demonstrated that their suggested architecture reduces operating expenditures by reducing the time it takes to execute services.

b) Reinforcement Learning for Smart City

Wu *et al.* [29] performed another study to provide a cognitive structure for IoT activities called Cognitive IoT (CIoT). The five cognitive tasks that can interact inside the framework are:

- The interpretation–action cycle.
- Huge data analysis.
- Semantic extraction and information exploration.
- Smart choice-making.
- Upon-request provisioning of services.

Both were deriving the semantics from analyzed data and discovering important patterns and rules as knowledge was highlighted as necessary for items in a cognitive context to comprehend and develop [30, 31].

Since the motion space is highly dimensional and continuous, applying RL methods [32] to the VF resource allocation problem is difficult. Previous RL investigations [33] explored practical issues with discrete or even Boolean solution areas, needing only a small number/type of actions (such as left and right for the traditional cart-pole problem). There is no upper restriction on the number of vehicles, services, or VFs in the VF resource allocation problem; hence, the action space is unbounded. In addition, the VFs should be able to take on only a fraction of the remote cloud's compute workload. The decision variables will thus be integers.

c) Cognitive Computing and Reinforcement Learning

Unsupervised and supervised learning are two categories of conventional machine learning techniques. In those techniques, the machines perform operations like regression, segmentation, and consolidation while the computers train the models using frequently fixed data. The machines can only receive a finite amount of data, though. Because they can only make predictions based on the information they have received, robots find it challenging to learn knowledge in nonlinear situations. However, various conditions can result in different tags for the same data, which means that different users can use the knowledge that computers have learned differently. Closed training with data input is the foundation of conventional and unsupervised learning. These conventional teaching techniques fall short of what is needed to increase machine intelligence over time. As a result, machine learning's hot research area now includes reinforcement learning [34].

The practice of reinforcement learning is very comparable to cognitive behavior. As an illustration, consider the situation of a child learning to speak. Typically, an adult may read a word aloud to a kid several times while pointing to the object it refers to or performing the activity it describes. The adult would correct the child if their understanding needed to be corrected due to poor judgment. When the child correctly answers, the adult will reward them. The surrounding environment plays a significant role in how well humans learn. By this time, reinforcement learning is used as an example, and it can learn from the environment and consider behavior. Several mechanisms for rewarding behavior have been established; for example, when behavior benefits an objective, it is rewarded, and when it does not, it is punished. There are several options available as we go towards the goal. As a result, each time a decision is made, it may not be the best, but it must be good for the machine to earn more money. Take AlphaGo [35] as an illustration. It uses reinforcement learning to play chess with itself after ingesting millions of chess matches for machine learning.

Each step taken throughout the self-learning stage may not be ideal, but it will likely result in the final victory according to the overall plan. The machine would experiment with novel routes in this process and rely on prior knowledge to maximize the desired return.

The improvisational play would be incorporated once the fundamental skills are understood, like when learning to draw. Data are produced during device efforts, but the ultimate goal is not regression, categorization, or grouping but maximum reward. Positive and negative attempts are significant for the device with this in mind. At every subsequent stage, it would learn from the mistakes made in prior tries. However, a machine's intellect is inadequate if it just converses with itself, similar to how it would be challenging for a child to learn to speak without interacting with others. Therefore, a learning system is not a good cognitive system if it implements its idea regardless of the external environment. A cognitive system should therefore communicate with humans directly.

Nonetheless, it would take a lot of time and labor if a person was specifically designated to communicate with a machine. Crowd-sourcing can make the device and expression of thoughts seem natural. The video game Foldit is a prime example [36]. In this game, a target protein is provided, and the players must put this protein piece by piece together using different amino acids.

The participants voluntarily take part in the amino acid assembly process. Moreover, if there are enough participants, the combined intelligence of this group of amateur players will be greater than that of a select group of professionals. This technique enables humans to subconsciously connect with machines through specialized cognitive computing software, hence increasing the intelligence of machines in specific application domains. Unconsciously, participants in crowd-sourcing supply a variety of information, which reduces the need for cognitive learning based on data and simultaneously introduces a novel approach to data processing.

2.4   Proposed Cognitive Computing RL Model

Here, we provide a comprehensive overview of an inherently adaptable and scalable cognitive computing system. This chapter uses data from various sensors to power several cognitive computing-based use cases. Our proposed architecture minimizes the effort to develop unique setups for each use case. Figure 2.2 depicts the proposed architecture. We can provide faster, real-time-based solutions to better serve the ever-changing information recorded by smart cities.

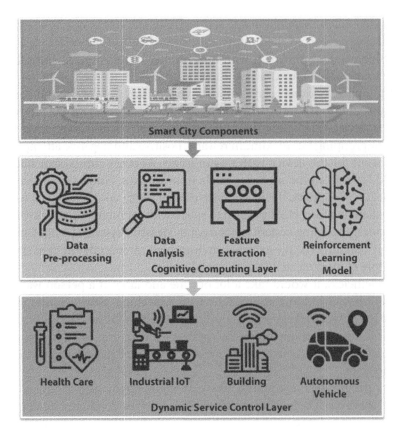

Figure 2.2 The Reinforcement learning model-based cognitive computing architecture for smart city.

The Smart city framework, the Internet of Things layer, the Data layer, the Cognitive computing layer, and the Service layer are the five pillars upon which the proposed computing architecture rests. The following section elaborates on creating a smart city with the help of Cognitive computing, a cloud-based client-server design.

a) Description of RL Enabled Cognitive Computing System
The designed cloud-based client-server structure is user-friendly and facilitates communication between many parties. The user can tailor their interactions with the server to their own needs. Data protection, confidentiality for users, data governance, and a protocol for exchanging data between users and service providers are all provided by this architecture. There is little doubt that others can use this framework to advance IoT research.

Network Operator, Centralized Server, and Services Seeker are the three primary nodes in the proposed system.

1) Service Provider: Raw environmental data are gathered by the Service Providers. The data collector could be a sensor, a smart gadget, or a citizen volunteer. The raw data can be sent to the server via these service providers. Inputs can be in the form of text, images, sound, or video.

2) Central Server: In this design, the server plays a vital role. The server will be the repository for all data. As was previously said, the server can accept information from numerous service providers. When the server receives data from the service providers, it will process it, keeping only the valid data and discarding the invalid or outlier data. The Service Seeker's request for service will be sent to the server. The server will respond to Service Seekers in real time whenever they send a request, regardless of the time of day.

3) Service Seeker: Those that seek out service are typically end users. They can query the central server for the city's current state at any time for data on a particular location and about events occurring in and around the city or photographs taken within the city.

b) Communication Network

The Network Operator, the Resource Seeker, and the Centralized Processor are the three primary parts of the proposed system. Several protocols have been established in their communications to limit the flow of information. Below are the rules they use to talk to one another:

Service Provider–Service Provider: There is no way for customers of different service providers to communicate. They are separate and distinct from one another. Data manipulation is risky if they can communicate with one another. This issue has prompted the implementation of new protocols.

Service Seeker–Service Seeker: Because they are completely oblivious to one another, Service Seekers cannot communicate.

Service Provider–Service Seeker: There is no way for the Service Provider and Service Seeker to contact one another. They are divided by a server in the middle. There is zero connection between the two.

Central Server–Service Provider: A constant two-way interaction exists between the Central Server and Service Provider. In most cases, the service provider will transmit data to the server. Still, if the server needs data from a specific city neighborhood, it will initiate contact with that neighborhood's service provider directly. The server can organize their groups to communicate with all the service providers with the same message more efficiently.

Central Server–Service Seeker: The Service Seeker will submit an inquiry for the requested services to the domain controller, and the host will respond in real time. That is why there is always two-way talk between the server and the person looking for service.

The communication layer is responsible for all types of transaction, in that Central Server–Service Provider plays a major role because they are the owner of data and computing systems. The Central Server–Service Seeker will establish only the transaction between them.

c) Cognitive Computing Layer

This layer specifies the steps used to develop the algorithm for cognitive computing. We emphasize the procedures of data collection, analysis, extraction of cognitive features, and machine learning. This layer generates an algorithm dependent on the selected features, allowing us to tailor our responses to individual smart city residents. We combine cognitive characteristics with machine learning models to create an AI system based on cognitive computing.

- Data preprocessing: A source's actual data may need to be revised or corrected. Artificial intelligence (AI) models based on cognitive computing that include this information will be less effective. If the missing information has no bearing on the conclusion, it can be omitted, and if necessary, it can be replaced.
- Data analysis: Data analysis is the process by which we decide which mental characteristics will be used to train our model. Instead of developing a unique cognitive model for each application in the smart city, multiple qualities that serve multiple applications are chosen. For instance, smart energy and farming systems can benefit from cognitive processes that combine cognitive and sensory understanding. Both connected devices and transportation systems can benefit from optimizing user-contextual live data and emotional states.
- Cognitive trait extraction: To train the artificial intelligence algorithm, we collect the necessary data in real time from

users via sensors spread across the smart city. Emotion, voice, location, and brain activity data are all collected from users to form a full picture of their mental state. User context and actions are used to glean user activity characteristics. Brain data, emotional states, and online interactions can all be mined to reveal personality traits.

- Machine learning: Because of machine learning, trained systems can analyze data and make predictions based on their learning. Due to advancements in artificial intelligence, machine learning software is now capable of independent pattern development and improved analysis. Artificial intelligence describes this development. Cognitive computing has advanced the field by fusing how humans and machines are trained to think more naturally. Algorithms that can autonomously acquire information and enhance it from big data sources like IoT devices are a typical application of machine learning and reinforcement learning.

d) Cognitive Enabled Artificial Intelligence

Training systems using supervised and unsupervised algorithms on data in a predetermined format are the backbone of conventional machine learning techniques. Due to its inability to acquire new knowledge independently, the gains produced by this approach are limited. Although machine learning has found success in many areas, like computer vision, analytical thinking, and cryptography, in the past, there are better fits for today's cognitive computing-based AI technologies. These solutions need more intelligence and customization of cognitive computing systems. Deep learning and reinforcement learning results suggest that they can be used to deploy the applications built on our cognitive computing architecture successfully. While deep learning can pick up a lot of surface-level features, reinforcement learning can pick up a lot of deep-level features [37].

The use of Deep Learning algorithms allows for the training of systems to exhibit human-like cognitive abilities. The left and right hemispheres of the human brain are each responsible for distinct mental processes. The brain's left hemisphere handles logical and rational thought, whereas the right hemisphere handles visual thought and the recognition and expression of emotions [38]. Cognitive computing technologies mimic human use of their reasoning and intuition faculties through data analysis in much the same way. Measuring an object is an example of logical and rational reasoning that can define it. To provide a relationship between input and output, perceptual reasoning will require the mapping of features.

Cognitive systems are more akin to human perception than conventional machine learning approaches since they can distinguish novel things without prior training.

Training systems to think like humans requires perceptual thinking, often known as feature mapping in the context of cognitive computing. Just using your mind's rational faculties won't cut it. Features like the position and size of eyes, nose, and mouth can be used to teach a computer to recognize a human face. Facial hair, unusual expressions, eyewear, and poor photography all make it harder for the system to identify a person's face [38]. The mapping relation between the person's picture and the outcome is derived intuitively, just as humans would. Using deep learning, we can classify images in a way that is eerily similar to how humans do it. Training a deep learning model requires a huge number of visual features, and the mapping data are employed in cognitive systems to generate unique answers.

In many ways, the behavior of reinforcement learning is similar to that of human learning. Reinforcement learning is a machine learning method that modifies behavior in response to rewards or incentives in the surrounding world. The system receives a reward if its chosen action is the best possible among several alternatives for achieving the goal. While the chosen strategy may benefit the system overall, there may be better options. The approach can be refined by continuous learning and development. Combining reinforcement learning with cognitive computing can improve agents' decision-making in complicated situations. Cognitive computing, for instance, may educate agents on the current condition of their surroundings and the best course of action to take to maximize their reward. It can also facilitate the process of experience-based learning and rapid adaptation for agents. Robotics, gaming, finance, and healthcare are just a few domains that can benefit from combining reinforcement learning with cognitive computing. Complex tasks, including item handling, navigation, and path planning, are all within the agent's capabilities in robotics. Agents can be taught to analyze market data and make predictions [39]. By allowing for more effective and efficient decision-making, the combination of reinforcement learning and cognitive computing has the potential to revolutionize numerous sectors.

e) Reinforcement Learning Model for Dynamic Control System

The reinforcement learning model is implemented in the cognitive system for dynamic control over real-time data. A single engine's worth of orders, supplies, and workload were analyzed, along with the current status of the model. The data showed that various environmental changes had taken place. The flexible production line that results from cognitive dynamic RL's control is fed into the system as an input for dynamic control planning. We looked at things like reduced total completion times, lower energy consumption, and more efficient machinery utilization to establish the worth of our solution. The proposed model was modified as precisely and accurately as possible with the help of utility value analysis. The intelligent agent reorganized the sweets production line in response to a change in road plans. Order quantity and planned decisions were kept apart in the RL structure for the packing line, even though the packing line's status and the intelligent agent's utility value were unaltered. These factors were considered when using the proposed architecture to estimate constant spaces. A streamlined version of high-monitoring data was constructed with fewer dimensions.

The proposed context-sensitive model was built for Raspberry Pi due to the difficulty of training RL models. The intelligent edge system typically takes in order, stock, and unit work data as inputs. The time it took, the amount of energy used, and how efficiently it was used all factored towards the final score. In our earlier work, the cognitive agents considered these aspects while deciding where to place the product range. The "intelligent actor" is always the sensor-equipped packing line. Edge Intelligent system helped us locate information that would allow us to pay attention to the office. The proposed model was trained offline using huge datasets from production clouds. The internet research and new information we gathered allowed us to refine the model.

2.5   Simulation Results

The Reinforcement method was used to evaluate both ML and Cognitive architecture for smart city on the prototype's underlying platform. The evolution platform is developed with the help of an Intel 5 CPU processor with 8 GB RAM and Raspberry Pi as the client node. The proposed model was implemented for trading of product in the smart city network. We pretended to be buyers by acting out their role in the supply chain through placing orders. We discovered that when the size of an order grew, the impact of the recommended mechanism planning on the assembly line also grew. As can be seen in Figure 2.3, we tracked the time it took to complete an order, the amount of energy the manufacturing line consumed, and the number of times each piece of machinery was utilized separately. Little order sizes made it difficult to distinguish between the three phases.

Figures 2.3 and 2.4 demonstrate that the static system and proposed technique outperformed the centralized scheduling system when the number of orders was increased to 2000. The proposed model provides better results regarding both time and energy use. As the number of orders increased, so did the efficiency of the proposed method. Figures 2.3 and 2.4 depict how long it takes to finish the job and the quantity of power consumed. Figures 2.3 and 2.4 illustrate that the proposed architecture had shorter time, energy, and utility changes as the number of tasks rose

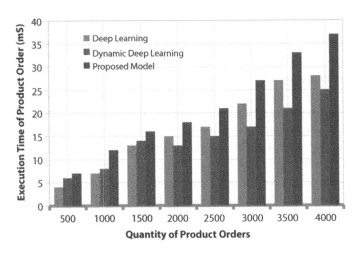

Figure 2.3 Evaluation time of the product order completion process.

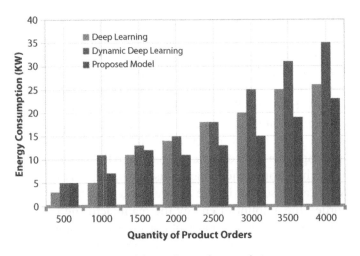

Figure 2.4 Energy consumption of the product order completion process.

compared to the other comparable models. From the evaluation results, we concluded that the combination of the cognitive and RL model was suitable for dynamic control of IoT devices.

2.6 Conclusion

This chapter presented an innovative computing architecture based on cognitive technology. The proposed architecture demonstrates how many cognitive qualities can be used to resolve different applications and offer real-time solutions on a platform for smart cities. There is less need for distinct settings for various dynamic smart-city-based applications. The proposed cognitive computing architecture addresses the present issue of complexity and scalability problems afflicting smart cities while managing big IoT data. We address related works in the context of cognitive computing and smart city architectures. We outline the system architecture for smart cities based on cognitive computing and reinforcement learning. The results of the experiments demonstrated that the proposed cognitive computing-based RL model was more efficient in terms of time, energy, and utility variances. In the future, we plan to implement dynamic control of edge computing devices using RL and cognitive computing models.

References

1. Lytras., M.D. and Visvizi., A., Artificial intelligence and cognitive computing: Methods, technologies, systems, applications and policy making. *Sustainability*, 13, 7, 3598, 2021.
2. Shenhar., A., Systems engineering management: A framework for the development of a multidisciplinary discipline. *IEEE Trans. Syst. Man Cybern.*, 24, 2, 327–332, 1994.
3. Sasikumar., A., Subramaniyaswamy., V., Jannali., R., Rao., V.S., Ravi., L., Design and area optimization of CMOS operational amplifier circuit using hybrid flower pollination algorithm for IoT end-node devices. *Microprocess. Microsyst.*, 93, 104610, 2022.
4. Wazid, M., Das, A.K., Bhat, V., Vasilakos, A.V., LAM-CIoT: Lightweight authentication mechanism in cloud-based IoT environment. *J. Netw. Comput. Appl.*, 150, 102496, 2020.
5. Sasikumar., A., Ravi., L., Kotecha., K., Saini., J.R., Varadarajan., V., Subramaniyaswamy., V., Sustainable smart industry: A secure and energy efficient consensus mechanism for artificial intelligence enabled industrial internet of things. *Comput. Intell. Neurosci.*, 2022, 1–12, 2022.

6. Sassi, M.S.H. and Fourati., L.C., Investigation on deep learning methods for privacy and security challenges of cognitive IoV, in: *2020 International Wireless Communications and Mobile Computing (IWCMC)*, vol. 714-720, 2020.
7. Ravi, L., Devarajan, M., Sangaiah, A.K., Wang, L., Subramaniyaswamy, V., An intelligent location recommender system utilising multi-agent induced cognitive behavioural model. *Enterp. Inf. Syst.*, 15, 10, 1376–1394, 2021.
8. Sasikumar, A., Vairavasundaram, S., Kotecha, K., Indragandhi, V., Ravi, L., Selvachandran, G., Abraham, A., Blockchain-based trust mechanism for digital twin empowered Industrial Internet of Things. *Future Gener. Comput. Syst.*, 141, 16–27, 2023.
9. Almeida, R., Oliveira, R., Sousa, D., Luis, M., Senna, C., Sargento, S., A multi-technology opportunistic platform for environmental data gathering on smart cities, in: *2017 IEEE Globecom Workshops (GC Wkshps)*, vol. 1–7, 2017.
10. Guevara, L. and AuatCheein, F., The role of 5G technologies: Challenges in smart cities and intelligent transportation systems. *Sustainability*, 12, 16, 6469, 2020.
11. Lamnatou, C., Chemisana, D., Cristofari, C., Smart grids and smart technologies in relation to photovoltaics, storage systems, buildings and the environment. *Renew. Energy*, 185, 1376–1391, 2022.
12. Nagarajan, M., Sasikumar, A., Muralidharan, D., Rajappa, M., Fixed point multi-bit approximate adder based convolutional neural network accelerator for digit classification inference. *J. Intell. Fuzzy Syst.*, 39, 6, 8521–8528, 2020.
13. Mouchili, M.N., Aljawarneh, S., Tchouati, W., Smart city data analysis, in: *Proceedings of the First International Conference on Data Science, E-Learning and Information Systems*, vol. 1-6, 2018.
14. Walter, A., Finger, R., Huber, R., Buchmann, N., Smart farming is key to developing sustainable agriculture. *Proc. Natl. Acad. Sci.*, 114, 24, 6148–6150, 2017.
15. Liu, Y., Wang, J., Chen, H., Cheng, D., Environmentally friendly hydrogel: A review of classification, preparation and application in agriculture. *Sci. Total Environ.*, 846, 157303, 2022.
16. Ali, H. and Choi, J.H., A review of underground pipeline leakage and sinkhole monitoring methods based on wireless sensor networking. *Sustainability*, 11, 15, 4007, 2019.
17. Siregar, B., Nasution, A.B.A., Fahmi, F., Integrated pollution monitoring system for smart city, in: *2016 International Conference on ICT For Smart Society (ICISS)*, pp. 49–52, 2016, July.
18. Fang, X., Misra, S., Xue, G., Yang, D., Smart grid—The new and improved power grid: A survey. *IEEE Commun. Surv. Tutorials*, 14, 4, 944–980, 2017.
19. Ghosn, S.B., Ranganathan, P., Salem, S., Tang, J., Loegering, D., Nygard, K.E., Agent-oriented designs for a self healing smart grid, in: *2010 First IEEE International Conference on Smart Grid Communications*, vol. 461–466, 2010.

20. Baig, M.M. and Gholamhosseini, H., Smart health monitoring systems: An overview of design and modeling. *J. Med. Syst.*, 1–14, 2013.

21. Sujith, A.V.L.N., Sajja, G.S., Mahalakshmi, V., Nuhmani, S., Prasanalakshmi, B., Systematic review of smart health monitoring using deep learning and artificial intelligence. *Neurosci. Inf.*, 2, 3, 100028, 2022.

22. Stojkoska, B.L.R. and Trivodaliev, K.V., A review of Internet of Things for smart home: Challenges and solutions. *J. Cleaner Prod.*, 140, 1454–1464, 2017.

23. Aleksic, S., A survey on optical technologies for IoT, smart industry, and smart infrastructures. *J. Sens. Actuator Netw.*, 8, 3, 47, 2019.

24. Asaithambi, S., Rajappa, M., Ravi, L., Optimization and control of CMOS analog integrated circuits for cyber-physical systems using hybrid grey wolf optimization algorithm. *J. Intell. Fuzzy Syst.*, 36, 5, 4235–4245, 2019.

25. Cellary, W., Smart governance for smart industries, in: *Proceedings of the 7th International Conference on Theory and Practice of Electronic Governance*, pp. 91–93, 2017.

26. Sasikumar, A., Senthilkumar, N., Subramaniyaswamy, V., Kotecha, K., Indragandhi, V., Ravi, L., An efficient, provably-secure DAG based consensus mechanism for industrial internet of things. *Int. J. Interact. Des. Manuf. (IJIDeM)*, 17, 5, 1–11, 2022.

27. Brown, B. and Laurier, E., The normal natural troubles of driving with GPS, in: *Proceedings of the SIGCHI Conference on Human Factors in Computing Systems*, pp. 1621–1630, 2012.

28. Wu, D., Lei, Y., He, M., Zhang, C., Ji, L., Deep reinforcement learning-based path control and optimization for unmanned ships. *Wireless Commun. Mobile Comput.*, 2022, 1–8, 2022.

29. Wu, Q. *et al.*, Cognitive Internet of Things: A new paradigm beyond connection. *IEEE Internet Things J.*, 1, 2, 129–43, 2014.

30. Mao, H., Alizadeh, M., Menache, I., Kandula, S., Resource management with deep reinforcement learning, in: *Proc. 15th ACM Workshop Hot Topics Netw.*, Atlanta, GA, USA, pp. 50–56, 2016.

31. Zhao, D., Chen, Y., Lv, L., Deep reinforcement learning with visual attention for vehicle classification. *IEEE Trans. Cogn. Dev. Syst.*, 9, 4, 356–367, Dec. 2017, doi: 10.1109/ TCDS.2016.2614675.

32. Zhu, Y., Mottaghi, R., Kolve, E., Lim, J.J., Gupta, A., Fei-Fei, L., Farhadi, A., Target-driven visual navigation in indoor scenes using deep reinforcement learning, in: *2017 IEEE International Conference on Robotics and Automation (ICRA)*, pp. 3357–3364, 2017.

33. Peng, X.B., Berseth, G., Yin, K., Van De Panne, M., DeepLoco: Dynamic locomotion skills using hierarchical deep reinforcement learning. *ACM Trans. Graph.*, 36, 4, 1–13, Jul. 2017.

34. Binder, W., Technology as (dis-) enchantment. alphago and the meaning-making of artificial intelligence. *Cult. Sociol.*, 1–24, 2022. 17499755221138720.

35. Milburn, C., Buse, K., Dhaliwal, R.S., Wills, M., Aldosari, R., Camarador, P., Siegel, J., Join the fold: Video games, science fiction, and the refolding of citizen science. *Des. Issues*, 39, 1, 70–87, 2023.

36. Sreedevi, A.G., Harshitha, T.N., Sugumaran, V., Shankar, P., Application of cognitive computing in healthcare, cybersecurity, big data and IoT: A literature review. *Inf. Process. Manage.*, 59, 2, 102888, 2022.

37. Middya, A., II, Nag, B., Roy, S., Deep learning based multimodal emotion recognition using model-level fusion of audio–visual modalities. *Knowle.-Based Syst.*, 244, 108580, 2022.

38. Kumar, T.A., Rajmohan, R., Pavithra, M., Ajagbe, S.A., Hodhod, R., Gaber, T., Automatic face mask detection system in public transportation in smart cities using IoT and deep learning. *Electronics*, 11, 6, 904, 2022.

39. Zhang, Y., Wu, C., Qiao, C., Sadek, A., Hulme, K.F., A cognitive computational model of driver warning response performance in connected vehicle systems. *IEEE Trans. Intell. Transp. Syst.*, 23, 9, 14790–14805, 2022.

3

Deep Recommender System for Optimizing Debt Collection Using Reinforcement Learning

Keerthana S.[1], Elakkiya R.[2*] and Santhi B.[1]

[1]*School of Computing, SASTRA Deemed University, Thanjavur, India*
[2]*Department of Computer Science, BITS Pilani, Dubai Campus, Dubai, United Arab Emirates*

Abstract

Unpaid debt strains the Field Collection Agent (FCA) and causes tremendous financial loss. Most research focused on analyzing the customer credit risk before loan approval to reduce the loss. Few studies focused on streamlining the debt collection process by minimizing the number of collection phone calls made to consumers and predicting the defaulters. None of the articles is focused on field collection agents' performance and priority allocation. We have framed the objectives such as analyzing FCA's performance, evaluating the customer's credit risk, and performing priority allocation by providing personalized recommendations. The FCA's performance is analyzed concerning different parameters such as collection percentage, number of visits, date and time, and visit intensity. Customer credit risk is analyzed with parameters such as CIBIL score, income, credit history, etc. FCA performance and customer risk categories are frequently changing. Our proposed Reinforcement Learning (RL) model adapts to these dynamic changes and prepares the allocation list, which maps the customer risk category to the corresponding FCA category. In addition, it gives personalized recommendations such as the statements, number of visits, date, time of visits, and the time interval between the visits. We have developed a Deep Recommender System for optimizing the Debt Collection Process (DRS-DCP). Our DRS-DCP reduces the human effort for dynamic priority allocation and provides personalized recommendations with a 76% hit ratio and 74% Normalized Discounted Cumulative Gain (NDCG).

Corresponding author: elakkiyaceg@gmail.com

Elakkiya, R. and Subramaniyaswamy V. (eds.) Cognitive Analytics and Reinforcement Learning: Theories, Techniques and Applications, (51–72) © 2024 Scrivener Publishing LLC

Keywords: Customer risk analysis, debt collection process, deep reinforcement learning, FCA performance analysis, priority allocation, recommendation systems

3.1   Introduction

The debt collection process is the method of collecting debt from customers. The pending dues have a significant impact on the finance sector. The revenue of the debt collection industry is improved by reducing unpaid debt. FCAs must be capable of collecting money from consumers. The FCA will attempt to contact the debtor by phone, email, or letter to request payment. Initially, debt collectors used phone calls to interact with customers to collect money from them. The FCA is required to collect the money with fewer phone calls. The authors implemented a method for optimizing the outbound calls made by debt collectors [1]. Only a few customers can respond to phone calls. Then, the debt collectors visit the customer's home to collect the debt. If a customer cannot pay, the debt collector offers partial payment to collect the money from the customer. The debt collectors cannot handle such customers if the overdue goes beyond several months. From the beginning, the customers are closely monitored and managed to prevent due. It is crucial to predict the default customers who miss the payment regularly. In the literature, the authors proposed a fusion method [2] for analyzing the default customers. In another work, the authors predicted the loan defaulter using Zhima's credit score [3]. The researchers proposed an ensemble method for predicting the defaulters [4]. The ensemble method is a combination of tree-based and Convolutional Neural Network models.

If the customer's overdue goes beyond several months, then that customer will be allocated the debt collection agency. The debt collection agency is a third party to collect money from the overdue customer on behalf of the FCA. An external debt collection agency is unnecessary if the FCA collects the money within the specified timeframe. Our method solves the problem of transferring customers to an external agency by improving the collection rate. Traditionally managers are responsible for the manual allocation of customers to the FCA, thereby specifying the instructions to collect money and how to handle every customer. The manager monitors the FCA performances and customer risk category to perform this mapping; this analysis takes time and human effort.

In this work, we have built the Deep Recommender system for optimizing the debt collection process with RL algorithms. We have used finance data from the private sector. Allocation is done to collect the money from the customers who missed the payments. In addition, managers give instructions to FCA. Instruction includes details such as when, where, and whom to collect the money from the customers. Our method reduces the human effort to map between the FCA and the customer by providing personalized recommendations. The deep Recommender system is built with RL and Deep Learning, called Deep Reinforcement Learning (DRL). DRL is used in various applications such as robotics, gaming, marketing, natural language processing, and autonomous vehicle. DRL is widely used in recommendation systems such as movies, products, news recommendations, etc. DRL-based recommendation systems provide good recommendation accuracy. DRL is suitable for a dynamic changing environment. In this work, FCA's customer risk category and performance keep changing. Depending on the nature of risk and performance, the model has to generate accurate recommendations and customer mapping. In our proposed DRS-DCP system, the debt collection process is optimized by using the following four objectives:

1) We have analyzed the performance of the FCA based on the collection percentage and concerned different parameters such as visit intensity, date and time of visits, and the gap between the visits to categorize the FCA performance.
2) We analyzed the customer's credit risk and payment behavior using DRL algorithms to predict the customer risk category.
3) Based on the FCA performance and customer risk category, we have allocated the customer to the FCA and recommended a suitable date, time, statements, and number of visits.
4) We have designed the debt collection process's state, action, policy, and reward.

In this chapter, Section 3.2 explains the basics and foundations of RL. Section 3.3 presents the different forms of RL. Section 3.4 gives the literature review of RL applications in finance and debt collection. The detailed methodology is given in section 3.5, and performance metrics are provided in section 3.6.

3.2   Terminologies in RL

Supervised Learning uses the pre-existing dataset, whereas RL does not use the pre-existing dataset. RL learns by exploring the environment randomly and utilizing the experience replay buffer to store the experience. The supervised and unsupervised algorithms have no exploration and exploitation, whereas RL tries to balance exploration and exploitation. RL performs sequential decision-making, whereas supervised and unsupervised algorithms make single decision-making. Feedback in RL is based on rewards and punishments. Labeled data are given as feedback in supervised learning.

RL solves the sequential decision-making process. Sequential decision-making is the process of making a succession of decisions over time in an environment that changes due to the decisions made. An agent in an environment selects the action based on the future expected reward. The value function represents the sum of all rewards from an initial state to the final goal state. The value is estimated through trial and error, where the agent communicates with its surroundings and updates its values based on the observed rewards. The agent selects a state action with the highest value function to get the expected long-term rewards. Bellman equation is a fundamental equation in RL and estimates the state value function. Model-based and model-free algorithms are the different types of RL. Model-based RL learns the environment's explicit model, which includes the transition function and reward. The agent aims to capture the environment's behavior and find the best policy. After learning the model, the agent may use it to simulate future trajectories and compute the predicted cumulative reward for various policies. The agent can then utilize these data to choose the best policy to maximize the expected cumulative reward. The model-based algorithm is best in terms of sample efficiency. The agent can learn directly from experience in model-free RL without an explicit environment model. Model-free can handle large and stochastic environments.

a) Markov Decision Process (MDP)

An RL process that satisfies the Markov Property (MP) [5] is the Markov Decision Process. The action at time t+1 creates a reaction in the environment, but the response may depend on everything that has occurred previously. Equation 3.1 [5] illustrates this dynamic.

$$\Pr\{R_{t+1} = r, S_{t+1} = s' \mid S_0, A_0, R_1 \dots S_{t-1}, A_{t-1}, R_t, S_t, A_t\} \qquad (3.1)$$

The probability of the reward at time t+1 is given by R_{t+1} and r is the reward for the particular State S and their corresponding action A. It indirectly depends on all possible past events $S_0, A_0, R_1, \ldots. S_{t-1}, A_{t-1}$.

If, on the other hand, the state signal has the Markov property, the environment's reaction at t+1 depends only on the state and action representations at t, in which case the environment's dynamics may be specified by providing only the state and action representations at t and is given in Equation 3.2 [5].

$$p(s',r|s,a) = \Pr\{R_{t+1} = r, S_{t+1} = s' \mid S_t, A_t\} \qquad (3.2)$$

If the environment meets the Markov property, one-step dynamics can only forecast the next state and reward based on the present state and action. The state collects all essential historical information. Once the state is recognized, history is discarded. These tuples <s, a, R, T, π> constitute the MDP.

b) State (s)

A state is when an agent observes the environment at a specific time. In RL, several states exist, including the initial, goal, and intermediate states. After performing a particular action, the RL agent transitions from one state to another. The state summarizes the information the agent needs to make decisions and conduct actions that maximize its cumulative reward over time. The state may be fully or partially observable in a given environment based on the available information.

c) Action (a)

The state transitions take place with action. The RL agent chooses the action from the policy. The agent's actions have a direct influence on the reward it gets as well as the state it transitions to. With a stochastic policy, the agent picks an action based on a probability distribution throughout the action space. With a deterministic policy, the agent always chooses the same action in each state. Stochastic policies are frequently employed in RL applications where exploration is crucial since they allow the agent to try multiple actions with some probability instead of always picking the same action.

d) Transition Probability (T)

The transition probability is the probability distribution of the future state given the current state and activity. In other words, it is the likelihood of changing from one state to another after doing a specific activity. Each state moves from one state to another with transition probability T(s, a, s'). s represents the current state and a corresponding action, while s' refers to the next state.

e) Policy (π)

An agent chooses an action based on the policy. Mapping of states to action π (s, a). Even a simple Lookup table or function can be a policy. Using policies, the RL agent learns to move from one state to another with the corresponding action. There are different policies in RL. In the deterministic policy, states and actions are already predetermined, but in stochastic policy, there may be many possible actions for a particular state.

f) Reward (R)

After performing each action, the RL agent will get a reward, i.e., either positive or negative. A positive reward is given to enhance the behavior of the RL agent. Similarly, a negative reward is given to minimize the particular behavior. The RL agent's long-term goal is to maximize the payoff. This reward plays a vital role in altering the policy. We can say how well or bad the action is based on immediate rewards, while the value function is needed in the long run.

g) Value Function (V)

The value function can be estimated with the help of reward. It is the sum of the reward collected from the initial to the goal state for a particular trajectory. A state, for example, may constantly provide a low immediate reward but have a high value because it is frequently followed by subsequent states that yield great rewards. The value function is calculated using the Bellman equation given in Equation 3.3 [5]. γ specifies the discount factor and ranges between 0 and 1. The optimal value function is calculated based on Equation 3.4.

$$v_\pi\left(s\right) = \sum_a \pi(a\,|\,s) \sum_{s',r} p(s',r\,|\,s,a)\left[r + \gamma v_\pi\left(s'\right)\right] \qquad (3.3)$$

$$v^*(s) = maxv_\pi(s) \qquad (3.4)$$

h) Model

The model specifies how the environment should act. For example, the model may predict the next state and reward given the current state and action. Models are used to plan and choose a course of action by studying prospective future events before they happen. In RL, there are two methods: model-based and model-free.

i) RL Algorithms

Q learning is based on a model-free RL that selects the optimal action based on the maximum Q value, and the value is calculated based on the obtained rewards. Q learning is suitable for the problem with fewer state and action spaces. SARSA is another model-free RL, but it does not utilize the maximum action value in the following state (like in Q learning). At the next stage, SARSA changes the action-value function with the actual action performed. DQN handles the large state and action spaces, and the Q value is estimated using a neural network. The actor–critic algorithm combines both value and policy-based. The actor selects the action, and the critic is responsible for evaluating the goodness of that action. The policy gradient algorithms are responsible for optimizing the neural network parameters to improve the algorithm's performance. Monte Carlo Tree Search (MCTS) is a model-based, on-policy algorithm that uses a tree search to explore the state space and pick the best action at each step. It has been effectively employed in games like Go and Chess.

On-policy and Off-policy are the two forms of learning approach in RL. On-policy uses the current policy, and the agent modifies it based on the experience it has obtained while following the same policy. On-policy indicates that the same policy governs the agent's exploration and exploitation. On-policy algorithms are often more conservative and take longer to converge, but they are more stable and have a lower chance of diverging. An off-policy learns by experience based on following the different policies. Off-policy is suitable for the problem with a fixed set of data. Q learning is an off-policy algorithm, whereas SARSA is an on-policy algorithm.

3.3   Different Forms of RL

a) Deep Reinforcement Learning

DRL combines deep learning and reinforcement learning to handle large state spaces and action sequences. Simultaneously, it enables the agent to learn more complicated features than older techniques. A neural network is used to estimate the value function. An agent interacts with its surroundings, learns from reward feedback, and makes decisions using Deep Neural Networks (DNN). DRL methods include Deep Q Network, Proximal Policy Optimization (PPO), Trust Region Policy Optimization (TRPO), Actor–Critic, DDPG, Twin Delayed DDPG (TD3), and soft actor–critic. DRL has been successfully used in various applications, including gaming, robotics, and natural language processing. DRL has been used to teach agents to play Atari games, control robots, and generate human-like language.

b) Multi-Agent Reinforcement Learning (MARL)

Two forms of MARL are the centralized and decentralized approaches. In a centralized system, one agent is responsible for coordinating all the agents in a system. In a decentralized system, each agent behaves independently and has a policy or value function. In MARL, each agent's purpose is to learn a policy that maximizes its predicted cumulative reward while accounting for the behavior of other agents in the environment. The state of the environment in MARL is often partly observable, which means that agents may not have access to all vital information about the environment. This can make learning optimum policies more difficult since agents must infer the intentions of other agents based on their observed behavior. MARL is used in various disciplines, including robots, multiplayer games, and social networks. MARL can improve coordination and decision-making in complex and dynamic contexts by learning optimum rules for groups of agents.

c) Inverse Reinforcement Learning (IRL)

IRL is a branch of RL that teaches a reward function that explains an expert agent's behavior. The agent is provided with a series of demonstrations in IRL, comprising a sequence of states and actions performed by the expert agent in a specific environment. IRL aims to deduce the underlying reward function that motivates the expert's conduct, which can then be utilized to train a new agent to execute the same job. The primary notion behind IRL is to assume that the expert's conduct is logical and driven by some underlying reward function that the agent is attempting to optimize. IRL aims to recover this reward function from the observed demonstrations by looking for a function that assigns greater values to states and actions chosen by the expert and lower values to states and actions not selected by the expert.

d) Hierarchical Reinforcement Learning (HRL)

The agent breaks the big task into several smaller tasks. It learns the policies at multiple levels of abstraction. HRL is formulated using Semi-Markov

Decision Process (SMDP). It is an extension of standard MDP. An agent takes action after observing the state, and once entered, it remains in the state for some time. The actions are considered extended. Actions can be learned through the policy or value-based method. HRL starts with simple tasks and gradually increases complexity to achieve the desired goal. HRL is used in robotics to find the optimal path and obstacle avoidance. While finding the path, it divides into several sub-problems to solve the problem. HRL learns the policies from the sub-task and combines them to solve the whole task. In gaming, HRL solves multiple sub-problem simultaneously and coordinates them to achieve the goal.

3.4 Related Works

a) Reinforcement Learning in Finance
In recent years, RL has been applied across multiple finance applications such as trading, credit risk analysis, and portfolio management [6]. Algorithmic trading is a typical use of RL in finance. RL algorithms may learn trading choices based on previous market data and feedback signals such as gains and losses. The authors proposed the trading strategy for multi-type data using the DQN algorithm [7] and extracted the features using Convolutional Neural Network (CNN) and Bi-directional Long Short Term Memory (BiLSTM). Traditional rule-based trading systems cannot adapt to changing market conditions. Traditionally, algorithmic trading is done by rule-based or Machine Learning algorithms. In a rule-based system, rules were framed manually, and decisions were made based on rules. ML algorithm was trained with past data to predict the future trading strategy.

In their work [8], the authors proposed the multi-agent DRL for trading to the dynamic nature. Each agent was responsible for trading at a particular time. The results were better than the single independent agents. In another work [9], the authors proposed hybrid RL for pair trading. They used techniques such as dimensionality reduction, dynamic replay, clustering, behavior cloning, and regression. RL is used in quantitative trading, and the model receives market circumstances as input and generates trading actions as output. RL can also be used to manage a portfolio. RL algorithms can learn to distribute the money among various investments depending on goals, such as maximizing returns while limiting risk. Portfolio management systems based on RL can adapt to changing market circumstances and rebalance portfolios automatically. The authors proposed the hierarchical Deep Deterministic Policy Gradient (DDPG) [10] for managing portfolios.

In their paper, the authors proposed a method for maximizing the portfolio return using RL [11]. They experimented with three portfolios: NYSE and NASDAQ market stock and index assets. The authors proposed SARSA and DQN for learning trading strategies [12]. Before passing the inputs to DQN, they performed feature extraction separately. The stocks, cryptocurrencies, and currency pairs were data used in their work. Value at risk, daily return, total return, and sharp ratio were the evaluation metrics. Table 3.1 gives the overview of the literature survey on finance applications implemented with RL.

Table 3.1 Literature review—RL in finance.

S. no.	Applications	References	Datasets	Algorithms
1	Trading	[7]	Yahoo Finance and S&P 500 stock market indexes	CNN, BiLSTM, and DQN
2		[8]	Forex (EUR/USD)	Multi-agent DRL
3		[9]	Yahoo Finance data	Hybrid (Twin Delayed Deep Deterministic Policy Gradient and Double Deep Q-network algorithms)
4		[12]	BTC-USD, GOOGL, KSS, AAPL	SARSA and Deep Q Network
5	Portfolio balancing	[10]	Yahoo Finance	Hierarchical DDPG
6		[11]	Portfolios from market index assets, NYSE and NASDAQ	RL with Long Short Term Memory (LSTM)

b) Debt Collection Process

The debt collection process is improved by optimizing the phone calls [1] made to the debtors by the debt collectors. They formulated the Markov Decision Process for scheduling the phone calls. The authors proposed the ML model [2] for predicting loan defaulters in their work. The fusion model, logistic regression, CatBoost, and Random forest were used. The lending club dataset was used to experiment and evaluate. The fusion model provides better accuracy than the three models. The loan defaulter was predicted using Zhima's credit score [3]. They used the Chinese dataset for analysis. The data imbalance was addressed by using the technique called Synthetic Minority Oversampling Technology (SMOTE). They analyzed the model with and without Zhima's credit score. Zhima's credit score improves the classification rate of defaulters. The author reviewed applications of Artificial Intelligence in finance applications [13], such as credit risk analysis, fraud detection, trading, loan underwriting, and customer behavior analysis. In their work, the authors compared traditional debt collection practices and the digital debt collection process [14]. The researchers proposed a hybrid ML model for predicting credit risk [15], which combines supervised and unsupervised algorithms.

In a conventional system, the debt collector communicates with customers to collect money, whereas a digital system sends automatic messages and emails. The debt advice officers provide financial advice in traditional methods, while digital systems offer the tools to interact with customers. Most of the customers pay the due on time. Some customers miss the payment schedule called non-performing loans. In their work [16], the authors suggested predicting non-performing loans using a machine learning approach. The European debt collection agency dataset was used in their work. They used 20 different types of machine learning to forecast non-performing loans. Random forests and boosted trees provide good forecasting performance. The authors proposed two-stage predictions in their work [17], such as loan rejection and default loan risk. Initially, they analyzed whether the loan was eligible for approval and later predicted the default rate for the approved loan. They used the lending club dataset. For the first phase, they used Support Vector Machine (SVM) and Logistic Regression (LR) algorithms, while in the second phase, Deep Neural Network was used along with LR and SVM.

The authors proposed a model [18] for predicting missed and purchase payments using LSTM for credit card holders. Real-time transactional and public non-transactional datasets were used in their work. The authors studied credit risk analysis using the machine learning algorithms such as SVM, k-Nearest Neighbor, Logistic regression, and Random Forest [19].

The Random Forest algorithm provides excellent accuracy. Personalized recommendations are based on the customer's past, present state, and long-term interactions to automate the decision-making process for debt collections. Their method combined behavior informatics, Deep Learning, Reinforcement Learning, and recommender systems to provide optimal decision-making. They proposed a reinforced coupled recurrent neural network to optimize decision-making [20]. In their study [21], the authors suggested financial institutions come up with better credit risk analysis to overcome consumer indebtedness. They also mentioned reasons for consumer indebtedness, such as financial literacy, emotions, materialism, and risk perception. In their work [22], the authors used the Bayesian network to predict the uncertainty in the debt collection process. They predicted the probability that the debtor would pay the total amount.

In another work [23], the authors proposed credit risk analysis using the DQN algorithm. They used a confusion matrix to adjust the reward dynamically. They used five datasets: The credit card fraud dataset and the German, Australian, Chongqing and Lending Club datasets. They framed two static and five dynamic reward functions. The evaluation metrics were accuracy, F1 score, True Positive Rate, and AUC. Setting the threshold for optimizing credit risk is essential in credit risk analysis. The authors [24] proposed an RL model for optimizing the credit risk threshold. They used Q learning with Monte Carlo simulation. Their model provides significant results compared to the existing methods. In their work, the authors proposed a method for predicting customer invoice payments based on past customer interactions [25]. They used both supervised and unsupervised algorithms. Logistic regression provides excellent classification accuracy. They predicted whether the customer would pay the amount on time without any outstanding debt. Table 3.2 gives the literature review on debt collection process using machine learning approach.

3.5   Proposed Methodology

Every day, the DRS-DCP model allocates the customers to FCA who have missed the due date. At first, the model analyzes the customer risk category and FCA's performance, as shown in Figure 3.1. Based on the analysis, the customer allocation is done. The RL model recommends statements, date, time, number of visits, and the time interval between visits. Initially, the model suggests the actions at random. After a series of explorations, the agent provides recommendations based on the maximum Q value. The Q value is calculated based on the reward average, and the reward

Table 3.2 Literature review—debt collection process.

S. no.	Applications	References	Work
1	Credit risk	[3]	Defaulter prediction using Zhima's credit score
2		[15]	Prediction of credit score using a hybrid model
3		[19]	ML-based credit risk analysis
4		[23]	DQN-based credit risk analysis
5		[24]	Threshold optimization for credit risk
6	Debt collection optimizations process	[1]	Optimize the phone calls to improve the debt collection process
7		[2]	Predicting the loan defaulters using ML models
8		[14]	Comparison between traditional and digital debt collection process
9		[16]	ML approach to forecasting the Non-performing loans
10		[17]	Prediction of loan acceptance and default risk
11		[20]	Automated decision-making for debt collection
12		[21]	Determination of consumer indebtedness using the partial least square technique
13		[22]	Used Bayesian network to predict the uncertainty in the debt collection process
14	Finance-related application	[13]	Reviewed the applications of financial services using Artificial Intelligence
15	Customer payment behavior analysis	[18]	Analyzing customer payment behavior using LSTM
16		[25]	Prediction of customer invoice payment

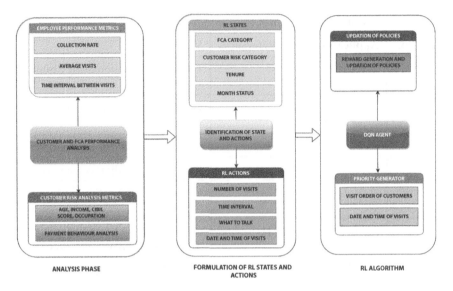

Figure 3.1 Debt collection optimization using RL.

is estimated based on the percentage collected from the customer follow-ing recommendations. The FCA's collection rate is calculated based on the daily and monthly targets. In a day, FCA is allocated 10–15 accounts for collections. FCA has to visit all the accounts assigned within the same day or the next day. By the end of the day, FCA has to update the status of the visit. Depending on the amount collected, the FCA updates the status, such as Fully Collected, Partially collected, and not paid. Sometimes, the customer gives the Promise To Pay (PTP) date. The RL agent generates the recommendations for the PTP date. In this way, the RL model generates the recommendations to the FCA until each customer's payment process is over.

a) Dataset

The dataset comprises four CSV files: collection, employee, visit, and cus-tomer data. We took this one-year data from the private financial sector. The collection details of each customer and their corresponding FCA allo-cated are available in the collection file. The employee data contain the details of the FCA, such as name, ID, monthly target, age, experience, etc. At the same time, the customer data includes the customer details such as name, age, income, CIBIL score, address, loan amount, tenure, and many more. Visit data contain the date, time of the visit, and collection status.

b) FCA Performance Analysis

The deep recommender systems use the DQN algorithm to recommend actions for improving the collection rate. FCA's monthly target and the overall amount collected are calculated from the dataset. The first phase is to analyze the FCA performance. The model categorized the FCA into different categories such as excellent, average, not so good, irregular, and insufficient data performers. Based on the collection percentage, the FCA is ranked from the highest to lowest. We have framed the deterministic rules for categorizing the FCA performances, which directly map the states to the actions. The rules are given as follows. The FCA should be in the top ranks for at least seven months; then, he is considered an excellent performer. If the FCA got the lowest rank for seven months out of 12 months, then he is regarded as a not-so-good performer. When the FCA exhibits rank variations at the top and bottom, the FCA is considered an average performer. If the FCA does not have seven months of data, then that FCA is considered insufficient. While the FCA does not provide consistent performance, he is regarded as an irregular performer.

c) Credit Risk Analysis

The second phase is to analyze the customer risk. The credit risk is analyzed before giving a loan to the customer. The initial stage in credit risk analysis is to collect information about the borrower, such as financial history, credit score, income, and assets. Credit reports, financial statements, tax filings, and other sources can provide this information. This information assists the lender in determining the borrower's creditworthiness and the amount of risk connected with the loan. We have framed the deterministic and stochastic policies by categorizing customers into high risk, medium risk, and low risk. Low-risk customers are less likely to miss payments, whereas high-risk customers consistently exceed the due date. The FCA has to approach the different categories of customers with a different strategy. Based on this two-phase analysis, the customer is allocated to the FCA. For example, a low-risk customer is assigned to the not-so-good performer to collect the dues.

d) Priority Allocation

The states and actions are framed concerning the debt collection process shown in Figure 3.1. FCA performance category, customer risk, tenure, and month status are considered the states. Actions are the number of visits, date, time of visit, the time interval between the visits, and what and how to talk. The DQN agent takes states as input. The neural network inside DQN estimates the Q value while the target Q value is calculated using the Bellman equation. Rewards are calculated based on the collection percentage. Initially, the model gives a random recommendation, such as date, time, gap, statements, and the number of visits, and analyzes the pattern of customer payment behavior.

Recommendations are provided based on this payment pattern analysis, occupation, date, and timing. For example, If the customer is a working professional, then the customer is visited on Sundays between 9 AM and 12 PM. Different timeslots are available that are suitable for a particular customer that timeslot is recommended. Gap specifies the number of days between each visit. If the customer has less overdue, the visit gap is greater, whereas for high-risk customers, the gap between the visit is less. Statements are recommended based on the customer risk category. If the customer had missed two months' payment, the statement recommended was "Ask them to give at least partial payments." Likewise, statements are generated depending on the customer status. The recommendation specifies the maximum number of visits. The FCA must collect money from clients within this limit. During the second recommendation, the agent has to recommend the date based on the PTP date. During every visit, the percentage collected is given as a reward to each FCA.

The Q table is maintained after each status update of FCA. The Q table consists of the customer name, FCA ID, customer risk category, FCA performance category, date and time of visits, the overall amount collected, number of visits, reward, statements, and Q value. The maximum Q value indicates that time, date, number of visits, gap, and statements are suitable for the collection process for the particular customer. After a few epochs, the model recommends the actions based on the maximum Q value from the Q table. The exploration and exploitation dilemma is solved by using the epsilon-greedy algorithm. We have set the epsilon with a value of 0.8, where it explores 80% of the time. The remaining 20% of the time, it performs exploitation.

3.6   Result Analysis

We have implemented this finance dataset in already existing algorithms in the literature, and the results are given in Table 3.3. We have considered the Hit ratio and NDCG as the performance measures. The Hit ratio is calculated based on the number of customers who have been successfully collected more than 70% of the amount divided by the total customer accounts assigned to the FCA. NDCG measures the relevance of the Top 1 recommendation.

Table 3.3 Comparison with existing works.

S. no.	References	Algorithms	Hit ratio	NDCG
1	[26]	Recurrent Neural Network with RL	70%	65%
2	[27]	Deep Hierarchical Category-based Recommender System (DHCRS)	69%	72%
3	**Proposed**		**76%**	**74%**

Table 3.4 RL algorithms—Result comparison.

S. no.	Algorithms	Accuracy	Recall	Precision	F1-score
1	Q learning	93.7%	0.91	0.96	0.93
2	SARSA	95.6%	0.95	0.91	0.93
3	Vanilla policy gradient	95.6%	0.90	0.88	0.90
4	**Proposed**	**97.8%**	**0.98**	**0.94**	**0.95**

We tried different RL algorithms for classification tasks (FCA performance and customer risk category), and their performance measures are given in Table 3.4. Among all the other algorithms, our proposed algorithm outperforms with a classification accuracy of 97.8%. We took the average accuracy of FCA performance and customer risk category.

a) Cumulative Reward
It is the sum of the total reward received for each episode. As the number of episodes increases, the cumulative reward increases. Our model is run for five months, and the corresponding reward obtained by the agent in each month is shown in Figure 3.2. It shows that up to the fourth month, its performance increases gradually and suddenly decreases after month 5. To perform well in real-world settings, the RL model must perform massive exploration, which takes time.

b) Episodic Time Step
The agent must complete the task with fewer time steps in an episode. The number of visits is considered as the time steps in an episode. In this

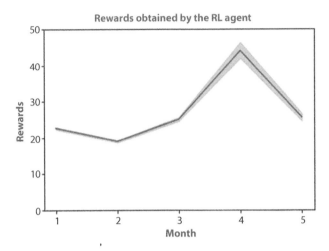

Figure 3.2 Performance of the RL model.

model, the episode ends when the FCA collects the total amount from the customers. The FCA mostly completes the payment from the customer in 4–5 visits. In extreme cases, the episode does not end for the customer who does not pay the money.

c) Learning Curve
A plot depicts the agent's performance (typically measured by cumulative reward) over time as it learns in the environment. This statistic can reveal information about the agent's learning progress and convergence. Figure 3.2 exhibits the learning curve and the upward trend up to the fourth month. The RL model continues to explore until it finds the optimal recommendations.

3.7 Conclusion

We developed the model for optimizing the debt collection process using DRL. It performs well in improving the collection rate and saving time. Field collection agent performance was analyzed concerning collection percentage during the analysis phase. The customer is allocated to the FCA based on their customer's risk and FCA performance category. Traditionally, managers do these analyses and allocations as it takes a lot of time and effort to understand each customer individually. Our model saves time and effort and provides excellent recommendations by improving the collection rate. Within this five-month execution, the DRL model provides a good hit ratio of 76% and an NDCG of 74%. In addition, it provides an excellent classification accuracy of 97.8%. As the model continues to update its knowledge, it suits a dynamic changing environment. In future work, we will develop an RL-based chatbot for optimizing the debt collection process.

References

1. Shoghi, A., Debt collection industry: Machine learning approach introduction. *J. Money Econ.*, 14, 4, 453–473, 2020.
2. Li, X., Ergu, D., Zhang, D., Qiu, D., Cai, Y., Ma, B., Prediction of loan default based on multi-model fusion. *Proc. Comput. Sci.*, 199, 757–764, 2021, doi: 10.1016/j.procs.2022.01.094.
3. Wang, H., Chen, W., Da, F., Zhima credit score in default prediction for personal loans. *Proc. Comput. Sci.*, 199, 1478–1482, 2021, doi: 10.1016/j.procs.2022.01.188.
4. He, H. and Fan, Y., A novel hybrid ensemble model based on tree-based method and deep learning method for default prediction. *Expert Syst. Appl.*, 176, 114899, January, 2021, doi: 10.1016/j.eswa.2021.114899.
5. Sutton, R.S. and Barto, A.G., Sutton Barto Second Book, vol. 258, MIT Press: Cambridge, MA, USA, 1998.
6. Singh, V., Chen, S.S., Singhania, M., Nanavati, B., kumar kar, A., Gupta, A., How are reinforcement learning and deep learning algorithms used for big data based decision making in financial industries–A review and research agenda. *Int. J. Inf. Manage. Data Insights*, 2, 2, 100094, 2022, doi: 10.1016/j.jjimei.2022.100094.
7. Liu, P., Zhang, Y., Bao, F., Yao, X., Zhang, C., Multi-type data fusion framework based on deep reinforcement learning for algorithmic trading. *Appl. Intell.*, 53, 2, 1683–1706, 2023, doi: 10.1007/s10489-022-03321-w.
8. Shavandi, A. and Khedmati, M., A multi-agent deep reinforcement learning framework for algorithmic trading in financial markets. *Expert Syst. Appl.*, 208, 118124, June, 2022, doi: 10.1016/j.eswa.2022.118124.
9. Kim, S.H., Park, D.Y., Lee, K.H., Hybrid deep reinforcement learning for pairs trading. *Appl. Sci.*, 12, 3, 944, 2022, doi: 10.3390/app12030944.
10. Wang, M. and Ku, H., Risk-sensitive policies for portfolio management. *Expert Syst. Appl.*, 198, 2022, 116807, July 2021, doi: 10.1016/j.eswa.2022.116807.
11. Lim, Q.Y.E., Cao, Q., Quek, C., Dynamic portfolio rebalancing through reinforcement learning. *Neural Comput. Appl.*, 34, 9, 7125–7139, 2022, doi: 10.1007/s00521-021-06853-3.

12. Taghian, M., Asadi, A., Safabakhsh, R., Learning financial asset-specific trading rules via deep reinforcement learning. *Expert Syst. Appl.*, 195, 2022, 116523, November 2020, doi: 10.1016/j.eswa.2022.116523.

13. Osterrieder, P.J., A primer on artificial intelligence and machine learning for the financial services industry. *SSRN*, 1–19, 2023.

14. Burton, D., Digital debt collection and ecologies of consumer over indebtedness. *Econ. Geogr.*, 96, 3, 244–265, 2020, doi: 10.1080/00130095.2020.1762486.

15. Machado, M.R. and Karray, S., Assessing credit risk of commercial customers using hybrid machine learning algorithms. *Expert Syst. Appl.*, 200, 116889, March, 2022, doi: 10.1016/j.eswa.2022.116889.

16. Bellotti, A., Brigo, D., Gambetti, P., Vrins, F., Forecasting recovery rates on non-performing loans with machine learning. *Int. J. Forecast.*, 37, 1, 428–444, 2021, doi: 10.1016/j.ijforecast.2020.06.009.

17. Turiel, J.D. and Aste, T., Peer-to-peer loan acceptance and default prediction with artificial intelligence. *R. Soc. Open Sci.*, 7, 6, 191649, 2020, doi: 10.1098/rsos.191649.

18. Ala'raj, M., Abbod, M.F., Majdalawieh, M., Jum'a, L., A deep learning model for behavioural credit scoring in banks. *Neural Comput. Appl.*, 34, 8, 5839–5866, 2022, doi: 10.1007/s00521-021-06695-z.

19. Tumuluru, P., Burra, L.R., Loukya, M., Bhavana, S., Csaibaba, H.M.H., Sunanda, N., Comparative analysis of customer loan approval prediction using machine learning algorithms. *Proc. 2nd Int. Conf. Artif. Intell. Smart Energy, ICAIS 2022*, pp. 349–353, 2022, doi: 10.1109/ ICAIS53314.2022.9742800.

20. Cao, L. and Zhu, C., Personalized next-best action recommendation with multi-party interaction learning for automated decision-making. *PloS One*, 17, 1, 1–22, January, 2022, doi: 10.1371/journal.pone.0263010.

21. Rahman, M., Azma, N., Masud, A.K., Ismail, Y., Determinants of indebtedness: Influence of behavioral and demographic factors. *Int. J. Financ. Stud.*, 8, 1, 8, Feb. 2020, doi: 10.3390/ijfs8010008, [Online]. Available: http://dx.doi.org/10.3390/ijfs8010008.

22. Köhler, B. and Fromm, H., *Predicting the Outcome of a Debt Collection Process Using Bayesian Networks*, vol. 7, 2022, [Online]. Available: https://hdl.handle.net/10125/79548.

23. Wang, Y., Jia, Y., Tian, Y., Xiao, J., Deep reinforcement learning with the confusion-matrix-based dynamic reward function for customer credit scoring. *Expert Syst. Appl.*, 200, 2022, 117013, June 2021, doi: 10.1016/j.eswa.2022.117013.

24. Herasymovych, M., Märka, K., Lukason, O., Using reinforcement learning to optimize the acceptance threshold of a credit scoring model. *Appl. Soft Comput. J.*, 84, 105697, 2019, doi: 10.1016/j.asoc.2019.105697.

25. Bahrami, M., Bozkaya, B., Balcisoy, S., Using behavioral analytics to predict customer invoice payment. *Big Data*, 8, 1, 1–13, 2020, doi: 10.1089/big.2018.0116.

26. Huang, L., Fu, M., Li, F., Qu, H., Liu, Y., Chen, W., A deep reinforcement learning based long-term recommender system. *Knowl.-Based Syst.*, 213, 106706, 2021, doi: 10.1016/j.knosys.2020.106706.

27. Fu, M., Agrawal, A., Irissappane, A.A., Zhang, J., Huang, L., Qu, H., Deep reinforcement learning framework for category-based item recommendation. *IEEE Trans. Cybern.*, 52, 11, 1–14, 2021, doi: 10.1109/ TCYB.2021.3089941.

Part II

COMPUTATIONAL INTELLIGENCE OF REINFORCEMENT LEARNING

4

Predicting Optimal Moves in Chess Board Using Artificial Intelligence

Thangaramya K.[1]*, Logeswari G.[2], Sudhakaran G.[3], Aadharsh R.[4], Bhuvaneshwar S.[4], Dheepakraaj R.[4] and Parasu Sunny[4]

[1]*School of Computer Science and Engineering (SCOPE), Vellore Institute of Technology, Vellore, India*
[2]*School of Computer Science and Engineering (SCOPE), Vellore Institute of Technology, Chennai, India*
[3]*School of Electronics Engineering (SENSE), Vellore Institute of Technology, Chennai, India*
[4]*Department of Computer Science and Engineering, College of Engineering Guindy, Anna University, Chennai, India*

Abstract

Artificial Intelligence (AI) has advanced to the point where computers have attained the grandmaster level in chess and are now competing for the World Championship. Despite numerous attempts using various AI techniques, chess has proven to be a highly complex challenge. As such, promoting chess as the primary testing ground for AI and highlighting its significant contributions to the field is crucial. Nowadays, artificial intelligence has grown to great heights and it has contributed a lot in the field of computer chess. However, it is computationally expensive and requires a lot of time to train. It is not feasible to implement such algorithms in the day-to-day machines we use. This brings in the need to develop a system that can work in any device without requiring high computation machines. This chapter proposes a new intelligent system for playing chess games in a computationally less expensive and more responsive way by extending the min–max heuristic search-based game playing algorithm with alpha and beta pruning, Convolutional Neural Networks (CNNs) with genetic optimization, and works in our day-to-day device without requiring high graphics, resolutions, and processors. In this model, the features of the game are included based on the game against a computer (with a different algorithm) and the game against another player (Multiplayer) along with a user-friendly Graphical User Interface

*Corresponding author: thangaramya112@mail.com

Elakkiya, R. and Subramaniyaswamy V. (eds.) Cognitive Analytics and Reinforcement Learning: Theories, Techniques and Applications, (75–102) © 2024 Scrivener Publishing LLC

(GUI). This work performs the User versus Artificial Intelligence task by using the Convolutional Neural Network (CNN) Algorithm with genetic optimization integrated into the min–max with the Alpha-Beta pruning algorithm for game playing and comparing using the metrics (ELO rating). It also displays the score at each step, which is used to analyze the corresponding player's position in the game. The main advantage of this model is that it is very user-friendly using GUI and also fast and responsive to the game design performed with minimal usage of the Central Processing Unit (CPU). From the evaluations carried out on this system, it is proved that the proposed system predicts the next move with more than 99% accuracy.

Keywords: Artificial intelligence, convolutional neural networks, genetic algorithms, alpha-beta pruning, ELO rating, central processing unit (CPU)

4.1   Introduction

A strategic board game played by two individuals, chess is also known as international or Western chess, different from other games like Xiangqi and Shogi. It has many applications as well including educational application, but is missing from the curriculum in schools (Vuk Uskoković 2023). The modern form of this game surfaced in the countries of Southern Europe after the 14th century. Presently, this game has become one among the most widely played games worldwide, with millions of enthusiasts engaging in the activity. Chess is a game of abstract strategy with no hidden information, and it is played on a board consisting of 64 squares made of an 8×8 grid. In this game, every player is provided with 16 pieces of chessmen at the start of the game, with one player commanding the white chessmen and the other player is responsible for commanding over the black chessmen. The pieces include one king, one queen, two rooks, two bishops, two knights, and eight pawns (Vuckovic 2008). The objective of this chess game is to provide a check to the king of the opposite team, which occurs when the king has no escape route, which can be implemented effectively using limited look-ahead (Arman Maesumi 2020). The game may also end in a draw under certain conditions. Early computer scientists sought to develop a chess-playing computer, and in 1997, the Deep Blue chess computer was able to defeat the World Champion of that time, namely, Garry Kasparov, in a match (Hsu Feng-Hsiung 2004, Campbell Murray, 1999). However, with the arrival of fast computers in recent years, they are able to play in such a way that the chess computers are more powerful than the best human players and have had a profound impact on the development of chess theory.

A chess-playing computer is made of powerful hardware and intelligent software and hence it becomes efficient in chess-playing games. This technology presents a unique opportunity for players to practice their skills, even when human opponents are not available. Additionally, computer chess provides various benefits, including analysis, entertainment, and training. Both hardware and software implementations of chess game use heuristic search algorithms for performing maximization of the score for one player and the minimization of the score for the other player using a static evaluation function for evaluating the scores more effectively. These systems use a function that generates plausible moves during the game playing mode, which uses the strategies that differ from those used by human players. The depth of the game trees can be incredibly large, consisting of a large number of nodes and hence the complexity of searching the depth of the tree is increasing (Pengsen Liu 2023). However, the present-day sophisticated hardware and Artificial Intelligence (AI)-based software make it possible to use the game playing algorithms and programs even today to play the chess game more efficiently with the opponent. Thus, chess software is mostly used now and it is more convenient.

Chess machines and programs come in a variety of forms, including stand-alone machines, which typically run a software chess program on a microprocessor, although sometimes they use some specialized hardware and rule-based intelligent software programs that will run on all types of computers available now. Most of the commercial chess-playing software and hardware that are available today are able to play at the strength of the super-grandmasters such as ELO 2700 or higher (Radek Pelánek 2016). The current chess-playing computers are made based on the multi-core technology with parallel computing capabilities-based CPU architectures. Moreover, the top chess-playing computer programs such as Stockfish are becoming more powerful than the human world chess champions in their ability to play chess. Typically, these programs consist of a chess engine that is connected to a graphical user interface (GUI) that allows the player to adjust playing levels, timing controls, and the other performance settings. Additionally, the GUI provides the player with features to set up the initial positions and also to edit the current positions and implementation of the reverse moves. Therefore, the current chess gaming systems offer the facilities like playing, winning, losing, and drawing the matches.

The chessboard has been designed as a square board consisting of an 8 by 8 matrix. Here, each square contains a piece object (with empty spaces represented by a "blank" piece). In addition, flag variables are used for the tracking of the movements related to the capturing and safeguarding of the queen and the king. In order to optimize the space and time during the execution of the min–max search algorithm (Elaine Rich *et al.* 2010), it is advised not to have separate boards at each branch of the tree. Instead, each plausible move must contain the information relevant to the pieces that are being moved from their starting and ending positions while making the plausible and optimal moves. This algorithm requires only one board object with the movements designed to maximize the score of the player. This score is used to make all the moves made during the search process. These representations keep track of individual pieces using separate instances, and bitwise operations can quickly provide information about the position of pawns and other pieces on the board. Researchers have spent years optimizing these representations for speed.

Computer chess has a rich and fascinating history that spans centuries. The first chess-playing machines, like the Turk, were actually hoaxes, designed to deceive people into believing that they were capable of playing chess. However, these machines sparked interest in the idea of building a true chess-playing machine, and this led to the development of modern computer chess. Chess has always been a game of strategic thinking, and this has made it a compelling challenge for academics and researchers. In the mid-20th century, pioneers like Alan Turing and Claude Shannon made significant breakthroughs in computer chess, laying the groundwork for future developments (Slate 1977). It is recognized that chess was a game of logical reasoning and decision-making, which could be broken down into a series of mathematical calculations.

As computers became more powerful, chess remained a fascinating problem for the developers to tackle. Over the years, chess engines have evolved from simple algorithms to Turing Machines, which can analyze millions of possible moves in a matter of seconds. Today, the best chess engines can outplay even the greatest human chess players, demonstrating the incredible progress that has been made in this field. Despite the many challenges that have been faced along the way, computer chess remains a compelling and rewarding area of research for computer scientists around the world. There is still much to be learned and discovered, and the ongoing development of new and more powerful chess engines is sure to lead to even greater achievements in the years to come. Deep Thought 2 was a computer program that was designed to play chess and it competed in several events from 1991 to 1995. Its predecessor, Deep Thought 1, had lost to a human grandmaster, which revealed some flaws in the program (Newborn 2012). To address these shortcomings, the developers created a new chip that had 8000 features, which was an improvement from the previous chip's 6400 features. This new chip enabled the program to combine search capabilities in C language with hardware search on the chip, resulting in the best possible moves being generated. Deep Blue, a later version of the program, was capable of analyzing up to 200 million moves per second, far surpassing a human grandmaster's ability to analyze only about three moves (Campbell Murray *et al.* 2002). This immense computational power allowed Deep Blue to beat the world champion Garry Kasparov in 1997, making it the first computer-based chess-playing program to defeat a human world champion in a given chess match (Hsu Feng-Hsiung 2004).

In recent years, artificial intelligence has made significant progress in solving complex problems, including chess. Machine learning techniques, such as Convolutional Neural Networks (CNNs), have been applied to chess, with promising results (Gajendran *et al.* 2023, Om Kumar *et al.* 2023). The CNN is a deep learning architecture consisting of convolution and pooling layers with a fully connected network that is capable of learning rules and applying them to classify data and handle the reasoning process based on the rules (Hassan *et al.* 2022). CNNs are well-suited to perform image recognition and classification tasks (Lee *et al.* 2021), which is why they have been successful in analyzing chess positions and predicting moves. Clark and Storkey (2014) developed a CNN-based model that achieved an accuracy of 44.4% in predicting professional moves in the strategy board game Go. This means that the model was able to predict almost half of the moves made by professional players in actual games, which is a promising result. These advancements in AI demonstrate the ongoing progress and potential of machine learning and other AI techniques in solving complex problems. In 2017, the computer chess world was stunned when Alpha Zero, a program developed by DeepMind, an AI research company that later became part of Google, beat Stockfish, a chess-playing engine. Alpha Zero used a different approach to chess than traditional engines like Stockfish. Instead of relying on human-crafted evaluation functions and searching through millions of possible moves, Alpha Zero used machine learning, specifically reinforcement learning, to teach itself how to play chess (McGrath *et al.* 2022). To train its neural networks, Alpha Zero played against itself many millions of times, starting with just the rules of the game as inputs. According to DeepMind, Alpha Zero played 44 million games of chess against itself in just the first 9 hours of training. Through this self-play, Alpha Zero discovered novel strategies and moves that were previously unseen in the history of chess, surprising even the world's top chess players. This breakthrough in computer chess demonstrated the power of machine learning to learn and improve in complex domains like games, without any human input or guidance beyond the initial rules. It also opened up new possibilities for using AI to solve other complex problems in fields such as healthcare, finance, and energy.

The min–max algorithm is the foundation of computer chess, enabling the search process to expand the search space by generating the next play of the tree (Vuckovic 2015). This algorithm uses the same function to minimize the score of the opponent's and also to maximize the score of the current player by using negation operation. The algorithm starts with the analysis of the root node and its children by expanding the tree and evaluating the plausible moves and then by backtracking whenever it is necessary. The values returned may be between −10 and 10 where the negative scores denote the chances of losing, a zero score denotes the chances of winning, and the positive scores denote the chances of winning. The Alpha-Beta pruning (Russel and Norvig 2010), is a commonly used pruning technique that significantly reduces the search space in the min–max chess-playing algorithm. It is used to keep track of both types of moves, namely, the best and worst moves for each of the players encountered up to now and it avoids the searching of the sub-trees that are going to provide the worse results. The alpha and beta cutoffs are integrated with the basic min–max algorithm without any loss of accuracy. With Alpha-Beta pruning, the search depth can be doubled without increasing the search time. To optimize the search, the possible moves at each node of the search tree are appropriately sorted based on their scores, which are evaluated by looking ahead one ply. While the intuitive sorting approach is to order moves from best to worst, it is not always the most efficient strategy. Since most chess moves involve small positional gains and losses rather than capturing pieces, it is better to order "stand pat" moves first. Thus, the moves are sorted using their scores, with the smaller ones coming first. With Alpha-Beta pruning and appropriate move sorting, the branching factor of the min–max algorithm used in chess-playing games will reduce from about 35 to around 25.

The Genetic Algorithm is a heuristic search algorithm that can be enhanced by incorporating analytical methods to optimize the board evaluation function. While some methods of evaluating a chess board are based on simple metrics such as the relative values of pieces, other factors like mobility, checking the opponent, and the benefits of protecting one's own pieces versus threatening the opponent's pieces, are not as easily determined intuitively. To address this issue, a Genetic Algorithm that evolves the evaluation function through successive generations can be used (Sreejith et al. 2022). This algorithm utilizes a tournament module that allows the computer to play against itself using different evaluation functions. Initially, the algorithm generates a set of random evaluation functions. These functions are then evaluated by pitting them against each other in a tournament. The top-performing functions are preserved, while the underperforming ones are eliminated. The tournament module is used in the algorithm, and it involves a group of 10 static evaluation functions that play against each other. The functions are then ranked based on their performance, and the top-ranked functions are selected for the next generation. To create the next generation, these top-ranked functions undergo mutation and crossover, which creates new evaluation functions with traits of the previous functions. The newly created evaluation functions are then tested against each other in a new tournament. The process of selection, mutation, and crossover is repeated for several generations until the desired level of optimization is achieved. By utilizing a Genetic Algorithm, we can obtain an analytical method of optimizing the board evaluation function that considers all relevant factors and leads to improved chess-playing performance.

In this chapter, a new intelligent game playing algorithm is proposed by extending the min–max algorithm with alpha-beta pruning by providing extensions using Convolution Neural Networks (CNNs) with feedback from genetic algorithms for weight adjustment and optimization. In this model, the min–max algorithm works in two types of plies, namely, the maximizing ply and the minimizing ply. This algorithm starts the search from the root node of the tree and goes up to the leaf nodes by applying depth first search with backtracking. For reducing the search space, alpha and beta cutoffs are applied to perform alpha-beta pruning. The algorithm uses two functions, namely, a static evaluation function and a move generation function that generates the plausible moves for the successful completion of the chess-playing game by player 1 by maximizing the scores for player 1 and minimizing the score for player 2 and vice versa appropriately. The moves are learned through training using the deep learning algorithm CNN, which is integrated with alpha-beta-based min–max algorithm and extended with genetic algorithm-based weight optimization.

To ensure fairness, the proposed algorithm provides a facility for making both sides get to play as both black and white. This algorithm then selects the top five performing functions and generates five new ones to replace the bottom five. This process is repeated for a desired number of iterations, with a default setting of 10. Two components of the algorithm were run during the experimentation where the "preservation" component preserved the top five functions while the "mutation" component kept one and mutated the other four. The mutations were done by pairing two from the best five function values. Moreover, the determination of a winner of the game is not always straightforward, as games are limited to 50 moves to accommodate time constraints and draws are possible. Additionally, when playing deterministically against itself with low plies, it is easy to do the same. To determine a winner, the algorithm computes the scores based on static evaluation function values. While the functions usually agree that the side that has more points is set to win, there is a small chance that each side will think it is winning in a close game due to the use of different evaluation functions.

The proposed chess game playing algorithm has been designed to be used in day-to-day machines. The level of play of the computer would be quite less and it would be much easier for grandmasters to defeat. Moreover, based on the experiments conducted in this work, it is found that the proposed system plays much better than 90% of the chess players. This game playing algorithm is a good contribution to perform an automated computer-based chess-playing game, and it also helped us to evaluate and compare the proposed algorithm with the other algorithms that used the machine learning based classification algorithms such as Naïve Bayes (NB) classifier, Decision Trees, and Support Vector Machines (SVMs). For the players, the proposed system assists to improve their gaming skill and their understanding of the position using the score displayed after every move. There is no big threat, but if the depth of alpha-beta algorithm or the number of layers is increased in the neural network, it may slow the game and consume a lot of Central Processing Units (CPUs). This proposed algorithm gives an idea of how computers understand chess and helps to understand how different algorithms work and how to improve its performance. It also educates the players about how computer chess game algorithms are evaluated based on ELO rating.

The implementation of this work was carried out using Python programming and the TensorFlow tool. The number of convolution layers used in this work is 5 and the number of max pooling layers was also set to 5. This model used the RELU function as the activation function for feature selection and the sigmoidal function was used as the activation function in the fully connected layer. Moreover, the number of hidden layers present in the fully connected network was set to 3. The training was performed with 80% of the benchmark dataset called Li-chess created using games with human experts and testing was carried out using 20% of the dataset. The rules learned from training with genetic CNN with alpha-beta min–max algorithm are stored in a knowledge base and they are applied when the games are played by the computers against the human champions. From the experiments carried out in this work, the classification and decision accuracy made by this proposed algorithm produced 99.5% and is able to win the human champions very tightly in lower levels and easily in the higher levels. The major contributions of this work are as follows:

1. The min–max heuristic search with alpha-beta pruning algorithm has been extended with CNN deep learning classifier to train the expert and intelligent moves to the game playing algorithm.
2. Genetic algorithms were used to perform the weight optimization in the fully connected layer and to make fast convergence in the convolution and max pooling layers of CNN.
3. The algorithm was tested with both benchmark datasets and human players.
4. The proposed model is able to play better than 90% of the human players who were experts in the game and hence it is an efficient model.

The rest of the chapter is organized as follows: Section 4.2 provides the literature survey. Section 4.3 depicts the architecture of the proposed work and explains the details of the proposed algorithm. Section 4.4 discusses the results obtained from this work and compares them with existing works. Section 4.5 gives conclusions on this work and shows some future works.

4.2   Literature Survey

Alpha Zero is a computer program developed by DeepMind, an AI and research company that is now part of Google (McGrath *et al.* 2022). In 2017, it made headlines by beating Stockfish, in a series of matches. What made Alpha Zero's victory so remarkable was the fact that it learned to play chess entirely on its own, without any human guidance. To achieve this, Alpha Zero used a technique called reinforcement learning. It was given only the basic rules of chess and played millions of games against itself to train its neural networks. These networks allowed Alpha Zero to evaluate board states much more quickly and accurately than other chess programs, including Stockfish. In fact, Alpha Zero was able to achieve the benchmarks that were required to win over Stockfish in 4 hours of training, according to DeepMind. This is because it did not need to analyze over 70 million state spaces per second, as Stockfish does. Instead, Alpha Zero relied on the neural networks to learn and also to make decisions on board states for making the correct moves. To run Alpha Zero, custom-made systems were required. These systems are often referred to as "Google Supercomputers" due to their powerful processing capabilities. In the case of Alpha Zero, it was reported by DeepMind that the program ran on four Tensor Processing Units (TPUs) during its matches against Stockfish.

In this paper, the researchers generalized this approach into a single algorithm called Alpha Zero, which can achieve superhuman performance in multiple challenging games, including chess, shogi, and Go. Again, the algorithm starts with no prior knowledge of the game except for the rules, and it learns entirely through self-play. In just 24 hours, Alpha Zero was able to achieve superhuman performance in all three games and convincingly defeated world-champion programs in each case.

While Alpha Zero is not available to the general public, its performance has inspired the development of many open-source neural network chess projects. These projects aim to create AI programs that can learn to play chess as well as Alpha Zero did, without any human input (David *et al.* 2016). Traditionally, computer programs that play games like chess, shogi, and Go have been developed using a combination of complex search techniques, customized adaptations for the specific game, and pre-programmed evaluation functions that are designed by human experts. These programs have been refined over many years to become the best they can be. However, a program called AlphaGo Zero, created by the AI research company DeepMind, provided superhuman-like performance while playing the game of Go using a different approach. Instead of relying on pre-programmed knowledge, AlphaGo Zero used reinforcement learning to teach itself on playing games against itself. This means that it learned to play the game entirely on its own, starting with no prior knowledge about the game except for the basic rules (McGrath 2022).

The research paper proposed a new approach to improve computer programs' ability to play Chinese Chess (Jiao *et al.* 2015). Traditional methods like Hill-Climbing and Simulated-Annealing have limitations, such as getting stuck in local optimum or converging too slowly. The Adaptive Genetic Algorithm (AGA) overcomes these limitations by using a genetic algorithm that adapts to the problem being solved. The result is a more powerful program that can play Chinese Chess better than previous methods. However, using a genetic algorithm like AGA is computationally expensive and time-consuming, which makes it less efficient compared to other algorithms. As a result, researchers are constantly looking for ways to improve the efficiency of algorithms to make them more practical for real-world applications.

One way to improve efficiency is to use parallel processing, which involves breaking up the problem into smaller parts that can be solved simultaneously on different processors or computers. This approach can significantly reduce the time it takes to solve a problem and can make genetic algorithms more practical for real-world applications. In addition to Chinese Chess, genetic algorithms were applied to a wide range of applications, such as optimizing industrial processes, designing new drugs, and developing neural networks for artificial intelligence (Panchal *et al.* 2021). The ongoing research and development of genetic algorithms and other optimization techniques continue to push the boundaries of what computers can accomplish.

Neural networks in AI are computer programs that are inspired by the structure and function of biological neural networks. These networks learn and adapt by processing various examples and patterns, without the need for specific programming for each task. With the use of image modulation techniques, the goal is to create an effective chess-playing software that can learn and play against a human opponent. The neural networks are trained using a benchmark dataset and they are tested using part of the dataset. In chess, neural networks have been used to develop chess engines that can analyze board positions and suggest moves. The use of neural networks in chess is still a relatively new field, and researchers are constantly exploring new techniques and approaches to improve their performance (Kumar *et al.* 2020, Om Kumar *et. al* 2023). However, the potential of this technology is enormous, and it is expected to lead to even more advanced and sophisticated chess engines in the future.

In the field of computer chess programming, to improve the search mechanism of chess programs by using co-evolution, many researchers worked on this area (McGrath *et al.* 2022, David *et al.* 2014, Christo *et al.* 2022, Gajendran *et al.*, 2023). Co-evolution is a process in which two or more species evolve together, influencing each other's evolutionary paths (Anisha Isaac *et al.* 2022). In the context of computer chess programming, this means evolving the chess program's search mechanism and the test suite used to evaluate its performance at the same time (McGrath *et al.* 2022). The paper highlights that the traditional methods of improving chess programs, such as brute force search, are not always effective and can be computationally expensive. Instead, the paper proposes using genetic algorithms, which are optimization algorithms inspired by biological evolution. The genetic algorithm works by starting with a population of randomly generated individuals and then iteratively selecting the fittest individuals to breed and create new offspring with variations. This process continues until the best individual, or solution, is found. This research paper's approach to co-evolution involves evolving the search mechanism of the chess program by using genetic algorithms, while simultaneously evolving a set of tactical test suites to evaluate the program's performance. The test suites are used to ensure that the program can correctly identify and respond to various tactical situations in the game of chess. However, the paper also acknowledges that there are drawbacks to this approach.

Overall, the paper highlights the potential benefits of using genetic algorithms and co-evolution to improve the search mechanism of computer chess programs. However, it also acknowledges the challenges involved in implementing this approach, particularly in terms of computation and training time proposed.

Many new techniques were proposed by different researchers (Lai 2015, Shalini Subramani *et al.* 2023, Milad Kohzadi Chegeni *et al.* 2023) to train computer programs to perform feature selection and classification and also to play chess using deep learning algorithms. At a high level, the technique involves creating a neural network that can analyze the current state of the chessboard and predict which moves are most likely to be successful. The network is trained on a large dataset of past games and uses these data to learn which moves tend to be effective in certain situations. The major advantage of the proposed system is that it is very effective at playing chess, even against strong human players. However, there are some disadvantages to this technique as well. One of the main drawbacks is that it requires heavy computing machines to process the vast amounts of data involved in training the neural network. This can make the process quite slow and resource-intensive. Another potential issue is that the network may not always make the best move in a given situation, since it is limited by the data it has been trained on. This means that it may struggle in situations where it has not seen enough examples to make an accurate prediction. Despite these challenges, this work provides new ideas in classification based on learning and provides accurate decisions on making moves in the chess-playing game. As computers continue to get more powerful and neural network algorithms improve, we may see even more advanced and effective chess-playing programs in the future.

In the research work by Vuckovic (2015), the author introduced a move generation model in chess-playing games by extending the min–max algorithm to improve the performance of the min–max algorithm in computer chess programs. At a high level, the min–max algorithm is used by the computer program to evaluate different potential moves in a game of chess. The program considers all possible moves, evaluates them using a scoring function, and selects the move with the highest score. However, this process can be very time-consuming, especially when evaluating complex game states. To speed up the process, the authors of the paper implemented a technique called the "Candidate Moves Method". This technique works by identifying which moves are most likely to be good candidates for evaluation, based on the current game state. By focusing on these candidate moves, the program can avoid wasting time evaluating moves that are unlikely to be successful. This work design implemented the technique in a distributed environment, which further improved the performance of the program. However, this approach requires heavy computing machines, as the program must evaluate a large number of potential moves and game states in order to make an informed decision. Despite these challenges, the "Candidate Moves Method" is an effective technique for improving the performance of min–max algorithms in chess programs. It is also a promising area of research, as techniques like this may be applied to other types of games and algorithms in the future.

In the paper by David *et al.* (2016), the authors introduced an extended neural network model based on a deep learning approach to train a chess-playing program using deep neural networks. At a high level, the technique involves training the program on a large dataset of several million chess games. The program uses these data to learn patterns and strategies that are effective in certain situations, without the need for any further domain-specific knowledge. One of the main advantages of this approach is that it can be very effective at playing chess, even against strong human players. However, there are some drawbacks as well. One of the main challenges is that it can take a lot of time to train the program on such a large dataset. Additionally, the program may exhibit a somewhat unpredictable or "adventurous" playing style, as it is not necessarily following a strict set of rules or strategies. As neural network algorithms continue to improve and more data become available, we may see even more advanced and effective chess-playing programs in the future. It is worth noting that "Deep Chess" is just one example of how deep neural networks can be applied to chess-playing programs. Other approaches, such as combining domain-specific knowledge with machine learning techniques, may also be effective in improving the performance of these programs.

The method proposed by Arman Maesumi (2020) introduced chess with limited look-ahead, which is a technique used by the authors to develop a chess engine that can make good moves in a game of chess with limited look-ahead. At a high level, the technique involves using a combination of search algorithms and machine learning techniques to evaluate potential moves in a game of chess. The program looks ahead a few moves in the game, evaluates the potential outcomes, and selects the move that is most likely to be successful. By developing more effective and efficient chess engines, researchers can help players improve their game and increase the level of competition in the sport. A new technique based on Artificial Intelligence (AI) was applied by Silver *et al.* (2018) to develop a chess-playing program using deep reinforcement learning. At a high level, the technique involves training a neural network to play chess through a combination of self-play and reinforcement learning. The program learns by playing against itself and adjusting its strategies based on the outcomes of those games. One of the main advantages of this approach is its generality—it can easily be adapted to play other zero-sum turn-based board games, such as Go or checkers. Additionally, it achieves state-of-the-art performance relatively quickly, without the need for a lot of pre-existing domain-specific knowledge. However, there are also some drawbacks to this approach. One major challenge is searching speed—the program may need to evaluate a large number of potential moves in order to make a decision, which can be a computationally expensive and time-consuming process.

Though many machine learning and deep learning-based algorithms are present in the literature on AI, only few works used the deep learning algorithms for move prediction. Moreover, the existing game playing systems take more time for convergence when they are employed for automatic feature selection and classification due to the complexity in weight optimization. The evolutionary algorithms such as Genetic Algorithms (Gas) are able to optimize the search algorithms and they reduce the search space. When GA is employed in a chess-playing game for optimizing the learning algorithm, it improves the decision-making process through fast and most accurate convergence and classification results. Therefore, genetic algorithms are used in this proposed work to support the CNN performance along with alpha-beta pruning based min–max algorithm.

4.3   Proposed System

Initially, the person starts the game and it launches the user-friendly GUI and goes into the game. The game has a main menu that asks if the user wants to play a game with another player or with AI. On choosing the option to play with another player, the game begins and each player's score will be displayed simultaneously. On choosing the Vs AI mode, the player gets to choose which algorithm he/she wants to play with, and simultaneously while playing, the user score will be displayed alongside.

The system shown in Figure 4.1 consists of components, namely, the benchmark dataset called Li-chess, alpha-beta pruning-based game playing algorithm, CNN model, and the proposed algorithm that uses GA for fast convergence. The game can be played against other users by choosing the Vs Human mode. Here, each player scores will be calculated and displayed simultaneously while playing the game. The player score will be calculated based on the coin and its position in the game. By choosing the Vs AI mode, the player gets to choose between three algorithms to play

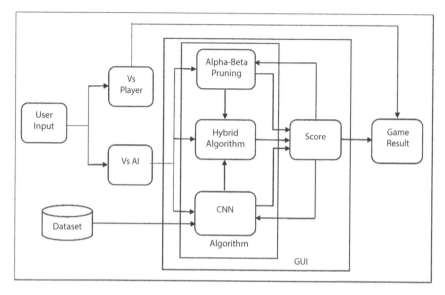

Figure 4.1 Proposed system.

against. By choosing the Vs CNN mode, the player gets to play against the CNN algorithm. By choosing the Vs Hybrid mode, the user gets to play with an algorithm which is a combination of CNN backed-up by Alpha-Beta Pruning algorithm. This mode offers quite a challenging game.

4.3.1   Human vs. Human

The functions used to build this chess game are the valid_moves—to list the valid moves of each coin, the is_check—to check if the king is under check, the search_moves—to list out the valid moves of each coin when under check, the pawn_promotion—pawn promoting into other higher rank coins, the Castling—castling, the active_player—player turn, the undo_moves—player can undo a move and go back to the previous state, and finally the en_passant—the rule permits the capture of a pawn that has moved two squares forward by an adjacent enemy pawn in a horizontal direction.

4.3.2   Human vs. Alpha-Beta Pruning

Alpha-beta pruning is an extended version of the min–max algorithm that optimizes the search space. Figure 4.2 shows the alpha-beta pruning process in the min–max search algorithm. Here, the alpha value of a max

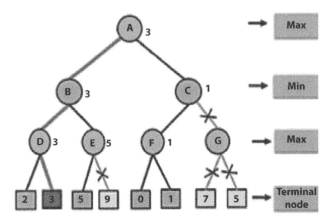

Figure 4.2 Alpha-beta pruning in min–max tree.

node is the minimum possible value for that node. Any node that generates a value less than this value is pruned. The beta value of a min node is the maximum possible value for that node. Any node that generates value more than this beta value is pruned. This process is called alpha-beta pruning. The human players need much time to think about the next plausible move. On the other hand, the min–max algorithm with alpha-beta pruning is able to reduce the search space by discarding a set of sub-trees from the search process.

Figure 4.2 shows a three-ply search tree in which the minimizing ply starts from the leaf nodes to their parent nodes, namely, D, E, F, and G. The maximizing ply from these nodes go to their parents, namely, B and C. The final minimizing ply ends with the node A, the root min node. Figure 4.3 explains the application of alpha-beta pruning in the chess-playing game. The move prediction process shown in Figure 4.3 consists

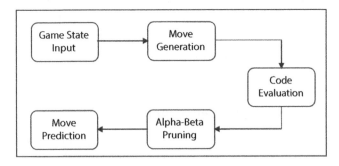

Figure 4.3 Architecture of move prediction using alpha-beta pruning.

Alpha-beta Pruning Pseudocode

```
FUNCTION findAlphaBeta(gs, plausible_moves, depth_of_tree, alpha_
value, beta_value, turn_multiplayer)
    SET global_variable bestMove
    IF depth EQUALS 0 THEN
        RETURN turn_multiplayer * scoreMaterial(gs)
    END IF
    SET maximum_score to -CHECK_MATE
    FOR EACH move IN plausible _moves DO
    gs.makeMove(move)
    SET bestMoves to gs.get plausible _moves ()
    SET score to -findAlphaBeta(gs, bestMoves, depth_of_tree - 1, -beta_
    value, -alpha_value, -turn_multiplayer)
    IF score > maximum_score THEN
        SET maximum_score to score
        IF depth_of_tree EQUALS DEPTH THEN
            SET bestMove to move
        END IF
    END IF
    gs.undoMove()
    IF maximum_score > alpha_value THEN
            SET alpha_value to maximum_score
    END IF
    IF alpha_value >= beta_value THEN BREAK
    END IF
    END FOR
    RETURN maximum_score
END FUNCTION
```

of five components, namely, Game State input, Move Generation, Code Evaluation, Alpha-Beta Pruning, and Move Prediction. Here, move prediction is performed using CNN with GA on Alpha-Beta pruning-based min–max algorithm, and it is shown in Algorithm 1 and Move prediction.

Figure 4.4 consists of six components, namely, the Li-Chess benchmark dataset, the pre-processing module, the training module, the prediction module, Game State input module, and the Move generation module that provides the most plausible moves to the player who uses it. Figure 4.5 shows the structure of the CNN used in this work. It has the input layer for taking the initial chess board positions using an element array of size 768. It has hidden layers

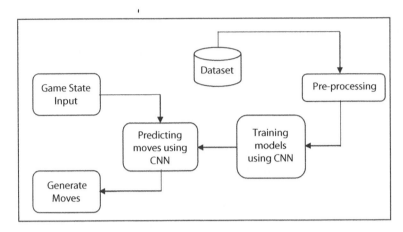

Figure 4.4 Architecture of predicting moves using CNN.

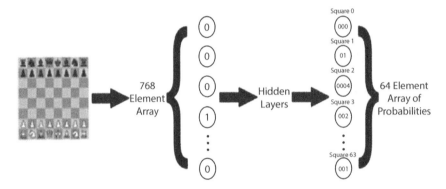

Figure 4.5 CNN layers.

called convolution and max pooling layers that provide the output to the output layer represented as 64 square values in the form of probabilities.

Neural Network Class

define method:
    set self.optimizer to 'Adam'
    set self.loss to 'categorical_crossentropy'
    create input_layer with shape (8, 8, 12)
    create x as Conv2D with 64 filters, kernel size 2, and strides (2, 2), using input_layer as input
    create x as Conv2D with 128 filters, kernel size 2, and strides (2, 2), using x as input
    create x as Conv2D with 256 filters, kernel size 2, and strides (2, 2), using x as input

flatten x

create x as Dense with 4096 units and 'softmax' activation, using flattened x as input

reshape x to (1, 64, 64) and assign it to output

create model with input_layer as input and output as output

compile model with optimizer=self.optimizer and loss=self.loss

assign model to self.model

train method:

  if EarlyStop is True:

      create es as EarlyStopping with monitor='loss'

  fit self.model with X and y for epochs and callbacks=[es] if EarlyStop is True

  save self.model as 'chess_model'

predict method:

  load 'chess_model' and assign it to model

  translate board using translate_board function and assign it to translated

  create move_matrix as model with translated reshaped to (1, 8, 8, 12)

  get legal moves using filter_legal_moves function with board and move_matrix as inputs, and assign it to move_matrix

  get the index of the maximum value in move_matrix using np.argmax with axis=None, and assign it to move

  convert move from an index tuple to a chess. Move object using np.unravel_index and chess.Move, and assign it to move

return move and 1

4.3.3   Human vs. Hybrid Algorithm

It is a combination of the two algorithms, the Alpha-Beta Pruning algorithm and Neural network. Here, it is implementing a neural network with three hidden layers and training with the chess game dataset that predicts moves using the dataset. In cases that are not explored yet, the moves are predicted with the help of the alpha-beta pruning algorithm of depth three. Figure 4.6 shows the process of the hybrid method used for predicting moves in chess.

Figure 4.6 shows the detailed architecture of the proposed system used for predicting the moves more accurately. It consists of eight components, namely, dataset, pre-processing module, training module, move prediction module based on CNN, alpha-beta pruning module, valid move generation module, generated moves, and the current input module. All these components work together to make the prediction of the next move more accurate.

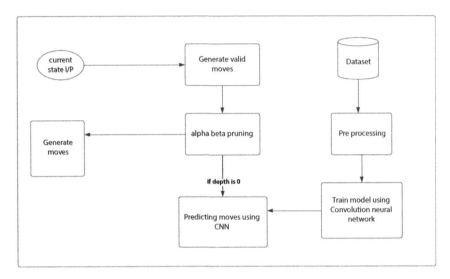

Figure 4.6 Architecture of predicting moves using the hybrid algorithm.

Hybrid Algorithm

```
def hybrid(gs, plausible _moves, depth_of_tree, alpha_value, beta_value,
turn_multiplayer):
global bestMove
side = 'White' if gs.whiteToMove else 'Black'
if depth == 0:
   board = chess.Board(boardToFen(gs, gs.board))
   engine = NeuralNetwork()
   flag = False
   move = engine.predict(board,side)
   for i in range(len(plausible_moves)):
         if str(plausible_moves [i].getChessNotation()) == str(move):
            gs.makeMove(plausible_moves [i])
            flag = True
            break
   t = turn_multiplayer * scoreMaterial(gs)
   if flag:
         gs.undoMove()
      return t
maximum_score = -CHECKMATE
for move in plausible_moves:
      gs.makeMove(move)
      bestMoves = gs.get plausible_moves ()
```

```
        score = -findAlphaBeta(gs, bestMoves, depth_of_tree - 1, -beta_
value, - alpha_value, -turn_multiplayer)
    if score > maximum_score:
            maximum_score = score
            if depth_of_tree == DEPTH:
            bestMove = move
    gs.undoMove()
    if maximum_score > alpha_value:
        alpha_value = maximum_score
    if alpha_value >= beta_value:
        break
return maximum_score
```

4.4 Results and Discussion

On entering the game, the user will need to choose the type of game that he/she wants to play as shown in Figure 4.7. It includes Human vs. Human or Human vs. Computer. On choosing the mode Vs AI, the user gets to choose to play against three different algorithms.

If the user presses "1" key—alpha-beta pruning. It shows the initial state and then the user goes to move the pawn. After moving the pawn, the opponent picks a move using alpha-beta pruning shown in Figure 4.8a. Figure 4.8b depicts playing against the Alpha-Beta Algorithm-driven system. From Figures 4.8a and b, it is observed that the game played against Convolution Neural Network is getting a better score than alpha-beta pruning. The game vs. the hybrid algorithm is also getting a score that is better than the above two algorithms. The normal player versus player

Figure 4.7 Main menu vs. AI Human vs. Computer (if the user presses key "1").

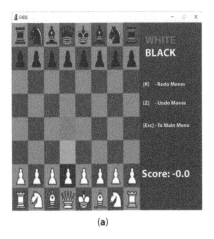 

(a)   (b)

Figure 4.8 (a) Alpha-beta, (b) Moving a pawn.

game is activated by choosing 2 in the main menu. Then, choose the player with special functions like castling and then move, as shown in Figure 4.9. The normal player is also getting high scores and intelligent moves. On winning a game, the respective score is depicted and it is shown in Figure 4.10.

From Figures 4.9 and 4.10, it is observed that castling and the use of deep learning algorithms with GA helps to make fast and accurate decisions to finish the game faster with success.

Figure 4.9 Choose castling.

Figure 4.10 Game over and get final score.

4.4.1 ELO Rating

Matches are conducted between the algorithms and initially have the ELO rating value of all as 1250. After every match, the ELO score is updated, and the ELO ratings are finally concluded with the respective algorithms shown in Table 4.1.

Table 4.1 ELO rating-based results.

Algorithms	Before		After updating using ELO Score		Result
	Rating 1	Rating 2	Rating 1	Rating 2	
Alpha-beta and CNN	1250 (Alpha-Beta)	1250 (CNN)	1250	1200	Alpha-Beta
Hybrid with CNN	1250 (Hybrid)	1200 (CNN)	1250	1180	Hybrid
Alpha-beta with Proposed Hybrid algorithm	1250	1250	1250	1250	Stalemate

From Table 4.1, it is proved that the proposed hybrid model with CNN and GA on the min–max algorithm with alpha-beta pruning provides higher rating than the existing works due to the use of CNN and GA.

4.4.2   Comparative Analysis

Figure 4.11 shows the move prediction accuracy analysis between the proposed algorithm and the existing algorithms.

From Figure 4.11, it is proved that the accuracy provided on the chess moves by the proposed algorithm is high compared to all other existing classification models including Naïve Bayes, ANN, and SVM algorithms. The performance improvement is higher in the proposed model due to the use of a deep CNN algorithm with genetic optimization for fast convergence and accurate prediction.

4.5   Conclusion

In this chapter, a new algorithm based on the min–max algorithm with alpha-beta pruning and the deep learning algorithm, namely, CNN-based classification, is proposed for predicting the next and optimal move for the chess-playing game. The use of alpha and beta cutoffs reduces the search space. Moreover, the deep learning algorithm is trained using expert player-based datasets and the rules learned during training are used to predict the next plausible move to be decided by the plausible move generator. The static evaluation function used in this work is supported further by the CNN-generated rules and hence the next moves are generated using rules

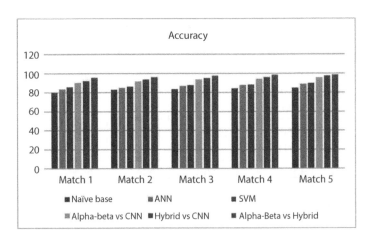

Figure 4.11 Accuracy analysis.

and inference mechanisms that are provided by the Artificial Intelligence-based decision-making. The results were evaluated for this work and also for the related work. Based on the experiments conducted in this work, it is proved that the proposed game playing algorithm enhances the prediction accuracy and reduces the time needed for decision-making while playing the chess game. Future works on this system can be the deployment of intelligent agents in the systems using the proposed algorithm for making fast and intelligent communications and decisions.

References

1. Maesumi, A., Playing chess with limited look ahead. *Artif. Intell.*, 2020. (Preprint).
2. Silver, D., Hubert, T., Schrittwieser, J., Antonoglou, I., Lai, M., Guez, A., Lanctot, M., Sifre, L., Kumaran, D., Graepel, T., Lillicrap, T., Simonyan, K., Hassabisa, D., A general reinforcement learning algorithm that masters Chess Shogi and Go through self-play. *Science*, 362, 6419, 1140–1144, 2018, doi: 10.1126/science.aar6404.
3. Isaac, A., Nehemiah, H.K., Dunston, S.D., Christo, V.E., Kannan, A., Feature selection using competitive coevolution of bio-inspired algorithms for the diagnosis of pulmonary emphysema. *Biomedical Signal Processing and Control*, 72, 103340, 2022.
4. David, O.E., Netanyahu, N.S., Wolf, L., DeepChess: End-to-end deep neural network for automatic learning in chess, in: *Artificial Neural Networks and Machine Learning – ICANN 2016*. Lecture Notes in Computer Science, vol. 9887, A. Villa, P. Masulli, A. Pons Rivero, (Eds.), Springer, Cham, 2016.
5. Lai, M., Giraffe: Using deep reinforcement learning to play chess. arXiv preprint arXiv:1509.01549, 2015.
6. Subramani, S. and Selvi, M., Multi-objective PSO based feature selection for intrusion detection in IoT based wireless sensor networks. *Optik*, 273, 170419, 2023.
7. Panchal, H., Mishra, S., Shrivastava, V., Chess moves prediction using deep learning neural networks. In *2021 International Conference on Advances in Computing and Communications (ICACC)*, pp. 1–6, IEEE, 2021.
8. Kumar, V., Singh, D., Bhardwaj, G., Bhatia, A., Application of neurological networks in an AI for chess game. *Research, Innovation, Knowledge Management and Technology Application for Business Sustainability (INBUSH)*, pp. 125–130, 2020, doi: 10.1109/INBUSH46973.2020.9392188.
9. Vuckovic, V., Candidate moves method implementation in MiniMax search procedure of the Achilles chess engine. *12th International Conference on Telecommunication in Modern Satellite, Cable and Broadcasting Services (TELSIKS)*, pp. 314–317, 2015, doi: 10.1109/TELSKS.2015.7357795.

10. Jiao, W., Yan-hong, L., Dong-ni, Q., xin-he, X., Adaptive genetic algorithm's implement on evaluation function in computer Chinese chess. *IEEE International Symposium on Communications and Information Technology*, pp. 1253–1256, 2015, doi: 10.1109/ISCIT.2005.1567095.
11. Thangaramya, K., Kulothungan, K., Logambigai, R., Selvi, M., Ganapathy, S., Kannan, A., Energy aware cluster and neuro-fuzzy based routing algorithm for wireless sensor networks in IoT. *Comput. Networks*, 151, 211–223, 2019.
12. Rich, E., Knight, K., Nair, S.B., Heuristic search techniques. *Artificial Intelligence*, 3rd ed. Tata McGraw Hill Education Pvt. Ltd, 52–76, 2010.
13. Russell, S. J. and Norvig, P., *Artificial intelligence: A modern approach*, Tata McGraw Hill, London, 2010.
14. Vuckovic, V., The compact chessboard representation. *ICGA J.*, 31, 3, 157–164, 2008.
15. Slate, D.J. and Atkin, L.R., Chess 4.5 - The northwestern university chess program. *Chess Skill Man Mach.*, 82–118, 1977.
16. Sreejith, S., Nehemiah, K.H., Kannan, A., A framework to classify clinical data using a genetic algorithm and artificial flora-optimized neural network. *Int. J. Swarm Intell. Res. (IJSIR)*, 13, 1, 1–22, 2022.
17. Gajendran, S., Manjula, D., Sugumaran, V., Hema, R., Extraction of knowledge graph of Covid-19 through mining of unstructured biomedical corpora. *Comput. Biol. Chem.*, 102, 107808, 2023.
18. Uskokovic, V., Natural sciences and chess: A romantic relationship missing from higher education curricula. *Heliyon*, 9, 4, 15015, 2023.
19. Campbell, M., Knowledge discovery in deep blue. *Commun. ACM*, 42, 11, 65–67, 1999.
20. Campbell M., H.A., Joseph, Hsu, F.-H., Deep blue. *Artif. Intell.*, Elsevier, 134, 1–2, 57–83, 2002.
21. Hsu, F. H., *Behind deep blue: Building the computer that defeated the world chess champion*, Princeton University Press, 2002.
22. Pelánek, R., Applications of the Elo rating system in adaptive educational systems. *Comput. Educ.*, 98, 169–179, 2016.
23. David, O.E., van den Herik, H.J., Koppel, M., Netanyahu, N.S., Genetic algorithms for evolving computer chess programs. *IEEE Trans. Evol. Comput.*, 18, 5, 779–789, Oct. 2014.
24. Chegeni, M.K., Rashno, A., Fadaei, S., Convolution-layer parameters optimization in convolutional neural networks. *Knowl.-Based Syst.*, 261, 110210, 2023.
25. Lee, S., Kim, J., Kang, H., Kang, D.Y., Park, J., Genetic algorithm based deep learning neural network structure and hyperparameter optimization. *Appl. Sci.*, 11, 2, 744, 2021.
26. Om Kumar, C.U., Gajendran, S., Balaji, V., Nhaveen, A., Sai Balakrishnan, S., Securing health care data through blockchain enabled collaborative machine learning. *Soft Comput.*, 27, 14, 9941–9954, 2023.
27. Hassan, M.R., Ismail, W.N., Chowdhury, A., Hossain, S., Huda, S., Hassan, M.M., A framework of genetic algorithm-based CNN on multi-access edge

computing for automated detection of COVID-19. *J. Supercomput.*, 78, 7, 10250–10274, 2022.

28. Christo, V.R.E., Nehemiah, H.K., Brighty, J., Kannan, A., Feature selection and instance selection from clinical datasets using co-operative co-evolution and classification using random forest. *IETE J. Res.*, 68, 4, 508–2521, 2022.

29. Gajendran, S., Manjula, D., Sugumaran, V., Character level and word level embedding with bidirectional LSTM–Dynamic recurrent neural network for biomedical named entity recognition from literature. *J. Biomed. Inf.*, 112, 103609, 2020.

30. Liu, P., Zhou, J., Lv, J., Exploring the first-move balance point of Go-Moku based on reinforcement learning and Monte Carlo tree search. *Knowl.-Based Syst.*, 261, 110207, 2023.

31. Om Kumar, C.U., Sudhakaran, G., Bhavadharini, R.M., Suguna, M., Krithiga, R., EHR privacy preservation using federated learning with DQRE-Scnet for healthcare application domains. *Knowl.-Based Syst.*, 275, 110638, 2023.

32. McGrath, T., Kapishnikov, A., Tomašev, N., Pearce, A., Wattenberg, M., Hassabis, D., Kim, B., Paquet, U., Kramnik, V., Acquisition of chess knowledge in alphazero. *Proc. Natl. Acad. Sci.*, 119, 4, 2206625119, 2022.

33. Newborn, M., *Kasparov versus deep blue: Computer chess comes of age*, Springer Science & Business Media, 2012.

5

Virtual Makeup Try-On System Using Cognitive Learning

Divija Sanapala and J. Angel Arul Jothi*

Department of Computer Science, Birla Institute of Technology and Science Pilani, Dubai Campus, Dubai, UAE

Abstract

Virtual try-on technology helps consumers to examine how specific things look on them before purchasing. This makes the buying experience easier, excitement-filled, and mess free. Customers are less likely to return items because they can test them before purchasing. This paper implements a virtual makeup try-on system using deep learning. The proposed system comprises two parts. The first section employs a proprietary convolutional neural network (CNN) to extract face key points from the input image. The second part aims to extract the desired facial regions and then apply makeup products on it using the mask and wrap up methods. The proposed model is used to apply lipstick, eyebrows, and eyeliner. The CNN model employed consists of four convolution layers and five fully connected layers. Batch normalization, max pooling, and dropout layers are utilized between the CNN's convolution layers. The Kaggle YouTube Faces Dataset is used in this work. For facial key point detection, the suggested model is compared to existing classical and deep learning algorithms. It is found that the proposed model has achieved an accuracy of 98.5%. It is user friendly and can help to increase market sales. It also helps users to test makeup virtually, saving time and money by avoiding product waste and post-application cleanup.

Keywords: Virtual, try-on, face landmark detection, convolutional neural network, facial key points

Corresponding author: angeljothi@dubai.bits-pilani.ac.in

Elakkiya, R. and Subramaniyaswamy V. (eds.) Cognitive Analytics and Reinforcement Learning: Theories, Techniques and Applications, (103–122) © 2024 Scrivener Publishing LLC

5.1   Introduction

For centuries, the most diverse people have used makeup and it is now a common item in many people's daily lives. Facial makeup can be used for a variety of purposes, including making someone appear more attractive, developing an artist's or model's personality, medical reasons, or simply for pleasure. It changes the skin's visual properties and texture. True face makeup experimentation, on the other hand, demands the use and disposal of several commodities, such as makeup, cleaning solutions, applicators, and so on [1]. Virtual try-on assists customers in making purchasing selections. They are free to experiment with several possibilities at their own pace until they reach the optimal decision. For example, users can mix and match different beauty products and experiment with different colored makeup products to determine what complements their present wardrobe and what suits them best. This technique lowers makeup product returns because customers can trial products before purchasing. Furthermore, consumers appreciate discussing their virtual try-on experiences since it is entertaining. This is also an excellent concept for brand marketing [17].

The virtual makeup try-on system has the following stages: (i) A facial image is used as an input to discover and identify feature points. (ii) Facial features such as the eyes, eyebrows, and lips are divided into cosmetic areas by feature points. (iii) The cosmetic color is applied to the facial components in the places designated for makeup. (iv) The face image is stitched together with the colored facial features [18]. This system is built with a deep learning convolutional neural network (CNN) model to recognize the facial important points of lips, eyes, and eyebrows for applying lipstick, eyeliner, and eyebrows, respectively. Deep learning models are developed using neural network architecture consisting of multiple layers and huge amounts of training data. These designs extract features from data directly, removing the need for manual feature extraction. Deep learning accelerates and simplifies processing. It is the foundation of recent AI developments and has the potential to continue to change a wide number of industries, allowing firms to obtain even more insights about the data [20].

The motivation of the paper is to develop a virtual makeup try-on system, which is a technologically advanced solution that allows people to virtually try on beauty products before purchasing them, which makes their shopping experience more engaging and convenient. A virtual makeup try-on system is developed using a CNN model to find facial features. The CNN model used in this model has an accuracy of 98.5%, which helps to provide reliable predictions of facial key points with a high degree of confidence. This means that the model can accurately apply makeup on the predicted facial key points. It helps to save resources and provides a competitive advantage. This can lead to a more satisfying user experience, leading to higher usage and loyalty.

The remainder of this work is organized as follows: we present a quick review of the literature in Section 5.2 and then the implementation on virtual makeup systems in Section 5.3, which includes model workflow, dataset description, image preprocessing steps, face landmark detection, and masking and wrapping techniques. Section 5.4 gives the details of experiments conducted and its findings. Then, the chapter is concluded in Section 5.5 with the conclusion and discussions.

5.2 Related Works

Ref. [2] published in 2019 describes an augmented reality system that allows users to apply virtual cosmetics to their faces while viewing a virtual reflection. A camera that captures Red, Green, Blue, and Depth data (RGBD) is utilized to identify and monitor face features, which are subsequently converted to a normalized 124-triangle 2D facial model. Additionally, the RGBD video stream detects finger touches on the face, which are then used to keep the cosmetics texture representation on the associated 2D facial triangle. Back projection of the cosmetics saved on the facial model to the camera image renders the face with virtual makeup. One of the model's flaws is the accuracy with which the AR system detects face features. Similarly, an article [3] in 2020 suggests a virtual cosmetics system that includes an offline-created database called golden samples that contains canonical patterns of facial components. The representative features of the facial components are then identified and the database is searched for the best-matching pattern. Users upload a face image to the internet for the facial landmarks to be retrieved. The Histogram of Oriented Gradients (HOG) and Support Vector Machine (SVM) approaches are used to find the face landmarks. The most suited pattern is then colored and stitched to the input image based on the user's preferences. This virtual makeup method produces good visual effects. One of the paper's downsides is the limited selection of beauty products accessible for virtual try-on.

Park et al. [4] aim to create virtual cosmetics using a person's particular color. Personal color or skin tone color matching is the process of selecting clothing and makeup colors that complement a person's skin tone, eye color, and hair color. This article describes a technique for making virtual cosmetics based on individual color choices. To do this, the user's own color is derived from the image first by studying the hues of the skin, hair, and eyes and then a cosmetic technique and palette are selected. The proposed method involves personal color analysis and a virtual makeup generator. It utilizes Dlib to extract facial landmarks, calculates precise coordinates for iris, hair, and skin, and applies makeup based on color analysis to the original input image. One of the paper's downsides is the possibility of the system making wrong color recommendations based on poor color analysis. Research [5] explained the importance of virtual makeup techniques for user convenience. One of the most crucial elements in the use of virtual cosmetics is accurate color reproduction. Therefore, it is necessary to reduce the hue discrepancy between the virtual and actual makeup outcomes. However, rather than concentrating on the accuracy of color reproduction, most earlier studies on virtual makeup recommended makeup styles. This article suggests a convolutional neural network-based method for faithfully recreating the color of lipstick. This study shows that, when compared to linear regression and multilayer perceptron (MLP) algorithms, the proposed method using a convolutional neural network produces the least amount of color difference. One of the downsides addressed in the research is the possibility of CNN making inaccurate color predictions.

The systems and methods for detecting makeup on an electronic device using an image of the user's preferred appearance as an input are described in Ref. [6]. The identified makeup is saved as virtual makeup in the relevant user profile and assigned to a virtual makeup library. Here, a Machine Learning (ML) model - CNN model is used. The strength, color, and style of the virtual makeup can all be changed by the user. Additionally, the user may be given one or more recommendations for virtual makeup depending on aspects of an image. The virtual makeup suggestions may appear on a screenshot, or they may appear live on the screen of the electronic gadget. To create the desired look, the user can change the recommendations for virtual makeup's strength, color, or style. One of the paper's downsides is the possibility of the system making inappropriate cosmetics recommendations based on insufficient or inaccurate makeup detection. The research [7] involves transferring the reference makeup to a face without makeup while preserving the reference face's makeup style and original appearance. This report systematically classifies cosmetics transfer technologies to comprehend the state of makeup transfer. Initially, it introduced and evaluated the conventional makeup transfer techniques in accordance with the method's development procedure. Specifically, the advantages and weaknesses of the makeup transfer methods based on deep learning framework are discussed. The discussion concludes with a few crucial points regarding the problems currently facing makeup transfer technology as well as its potential for future growth.

In a research [8], a three-stage design method was suggested: a straight-forward and practical virtual try-on model based on photographs so that the texture in the fitting results may be accurately mapped in the context of intricate figure posture and background. The final step was to style and reduce the fitting results to create pleasing outcomes. To enhance the detail quality of image synthesis, this work introduces the innovation of the pix22Dsurf-Generative Adversarial Network (GAN) method. The strategy described in this research can deliver better results, as demonstrated by both the quantitative evaluation and the user perception opinion. One of the disadvantages mentioned in the research is the possibility of the cosmetics style transfer resulting in unrealistic or unnatural-looking makeup. The Characteristic-Preserving Virtual Try-On Network (CP-VTON) pipeline was proposed in a study published in [9], which is a completely learnable image-based virtual try-on pipeline for the development of characteristic preserving images and consists of a new geometric matching module and a try-on module with a new merging technique. The geometric matching module attempts to align the target person's physique with in-store garments using significant spatial displacement. The try-on module learns to keep the qualities of properly aligned clothing and makes it appear more natural, which was lacking in previous models. Several tests have shown that the entire CP-VTON pipeline generates high-fidelity virtual try-on results that adequately replicate essential aspects of in-store clothing. One of the limitations addressed in the research is the possibility of the system causing artifacts or distortions in the virtual try-on image.

In Ref. [10], a novel image-based virtual try-on method—Outfit-Virtual Try On Network (VITON)—makes it easier to see how a combination of clothing pieces chosen from a variety of reference photographs comes together to make a cohesive outfit on a subject in a query image. The algorithm used in this paper has two unique characteristics: it is cheap because all that is needed is a sizable collection of single photographs of people wearing diverse clothing without explicit 3D information, both actual and catalog, and the second characteristic is that it allows control over the sort of clothing rendered in the finished outfit and can synthesize images of several garments that have been combined into a single, cohesive outfit. This method creates a segmentation map with geometric accuracy that modifies the shape of the chosen reference clothing to fit the target person. The potential for the method to produce unnatural-looking garment wrinkles or distortions is one of the downsides addressed in the research. Han et al. [11] demonstrate VITON, which uses only RGB photos to move apparel from a product image to a person. A multi-task encoder–decoder trained on a detailed, clothing-neutral person representation generates a coarse sample. With the help of a refinement network that learns the ideal composition, the rough findings are further improved. Experiments were done on a recently compiled dataset and both numerically and qualitatively promising findings were obtained. This suggests that the pipeline used in this paper for 2D image-based synthesis can be utilized instead of more expensive 3D-based techniques. This technique makes the garments look more natural on the people, which was a concern not highlighted in the previous research. One potential disadvantage of the suggested method is its dependency on clothing item and body segmentation.

The multi-pose guided virtual try-on system, according to research [12], enables clothes to be put on an image of a person while they are in various poses. This study suggests a Multi-Pose Guided Virtual Try-On Network (MG-VTON), which creates a new person image from an input image by altering human poses and including apparel. In addition to a human parsing model to guide image synthesis, a Wrapped Generative Adversarial Network (Warp-GAN) that learns to produce realistic images by correcting misalignment caused by various stances, and a refinement render that restores texture details, MG-VTON divides virtual try-on work into three stages. For the multi-pose guided virtual try-on challenge, a new dataset was created including images of people in a greater variety of positions and clothing. Experiments demonstrate that MG-VTON significantly outperforms both currently employed methods. One of the main drawbacks of the suggested method is the requirement for exact and precise position estimation. According to Gustavsson *et al.* [13] in 2022, a comprehensive framework was presented for manipulating cloth that is based on landmark identification and category categorization. The network was trained for garment categorization and landmark recognition for robotic manipulation application using a sizable publicly available dataset of fashion image data and a data augmentation technique termed elastic warping. They carefully assessed the network's performance as well as the consequences of the elastic warping. This article demonstrated how the method's parameter can be adjusted to meet a certain target distribution. The authors acknowledge that the availability of high-quality input photos limits this method.

Similarly, Ziegler *et al.* [14] explained the utilization of information and tools created for fashion image analysis in challenges involving vision-based robotic garment manipulation. This paper focused on methods that generalize from massive fashion datasets to smaller, less structured information gathered in robotic labs. To improve the model's generalizability, training data augmentation techniques like elastic warping and model modifications like rotation invariant convolutions were used. Tests show that, when applied to previously unexplored datasets, this method outperforms state-of-the-art models in terms of apparel category categorization and fashion landmark recognition. For accurate detection and classification, one drawback is the requirement for high-quality input photos with clear and visible landmarks. A mobile-based virtual try-on system that overcomes the issues of high cost, conflicting computing complexity, and simulation effects is described in the work [15]. This work consists of a number of modules, including automatic 3D face reconstruction from a single image, real-time local clothing simulation, and auto-skinning, which do not require the use of clear, high-quality photographs. According to the experiments, the virtual try-on system described in this paper can provide better-fitting results while necessitating less expensive architectural and computational requirements, leading to a favorable virtual try-on system experience when utilizing a mobile device. The precision of the system's clothing segmentation algorithm, which affects the caliber of the virtual try-on experience, is one potential restriction.

Similarly, another paper [16] suggests improvements in managing difficult postures of the input person photos as compared to the prior approaches. Separate components should be bent independently to mimic the occurrences of overlap and, due to the diverse natures, the movement of various body parts. A revolutionary virtual try-on method is put forth in this study. They provided a geometric feature-based warping technique to distort the sleeve that resolves several problems with earlier approaches. This warping process is supervised by using the body's constraints and key poses motions. Two learning-based courses are used that facilitate more effective warp computation and synthesizing while taking care of the occluded source and achieve a seamless output garment area. Even though using this technique results in performance but not just parsing and landmark calculation increases overhead potential for directing appearance flow-based techniques. One drawback of this system is that it needs high computational power, which is an issue to run on devices with low processing power. However, the present algorithms are slow, increase the overall cost of the system, and have significantly lesser accuracy. To save time and work often expended on executing this procedure manually, the paper offers an algorithm that helps users to virtually try on makeup. The algorithm described in the study aims at cutting down on both the system's overall cost and computational time. It also tries to improve the system design's correctness [19].

The main gap that is addressed in this paper is the lack of clarity of makeup applied in the previous virtual makeup try-on models and the accuracy has been increased in this model when compared to other traditional deep learning and pre-trained models. Increasing a model's accuracy

Table 5.1 Literature survey table summary.

Ref.	Aim	Methodology	Metrics
[2]	Virtually makeup using augmented reality	Three key processing components: Facial Touch Event Manager, Face Information Manager and Virtual Makeup Manager	Touch detection accuracy: 2.2 mm
[3]	Virtual cosmetics system using golden samples	Facial landmark detector is employed with feature representation.	Favorable visual outcomes for users' makeup trials.
[4]	Virtual makeup based on personal color analysis	Dlib Facial Landmark detector	Accuracy of personal color feature extraction: iris – 89.54%, hair – 86.23% and skin – 93.34%
[5]	Predict hues of lipstick	Convolutional Neural Networks	The color difference noted using CNN is less than 10 when compared to Multi-Layer Perceptron (MLP) and regression which has a difference greater than 10.
[6]	Virtual makeup using makeup detection and recommendation	ML model - CNN	Virtual makeup applied using electronic devices for users to achieve their preferred makeup appearance
[7]	Makeup style transfer	Deep Learning	Average time – 184.30s and covariance – 6.32
[8]	Virtual try-ons in apparel industry.	Generative adversarial networks.	Structural similarity – 0.859

(Continued)

Table 5.1 Literature survey table summary. (*Continued*)

Ref.	Aim	Methodology	Metrics
[11]	Virtual Try-On Network to transport a piece of clothing from a product image to a human	Step1: multi-person encoder decoder system Step 2: Refinement network. VGG19 has been used here	Accuracy: 77.2%.
[12]	Multi-pose Guided Virtual Try-on Network	Warping Generative Adversarial Network (Warp-GAN)	Structural similarity – 0.744
[16]	Skeleton-based Features in Virtual Try-On	Two learning-based models: synthesizer network and a mask prediction network.	Frechet Inception Distance score (FID) – 15.74

is an important goal because it can improve model performance and efficiency and lead to new discoveries and insights in the area. The clarity of the image is also very important for the users to get a good idea of how makeup looks on them and makes the makeup look more realistic and natural. Table 5.1 summarizes work on literature review.

5.3   Proposed Method

This research work has two parts—the first one is getting the facial key points from the input image and the second part is to use these points for applying makeup using mask and wrap up methods.

Figure 5.1 shows the architectural diagram of the proposed method. The input is taken either from a webcam or an image. After preprocessing the input image, face landmark detection algorithm is used to determine the facial key points of lips, eyebrows, and eyes. A mask is applied on the facial region to extract the necessary facial part. The color is then wrapped on the masked image and placed on the original image, which is the final image after applying makeup.

a) Input Data
The dataset used in this work is from Kaggle YouTube Faces Dataset [21]. This dataset consists of images and facial key points. The dataset consists

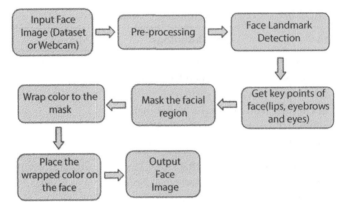

Figure 5.1 Model workflow.

of 5768 images of 1595 different individuals. The image resolution varies, with most of them being 320 × 240 pixels. The train test split ratio is 60:40 such that the training images consist of 3462 images and testing images consist of 2306 images of faces from YouTube videos of people with different facial expressions to correctly predict the key points. The training key points csv file consists of 68 x and y (136 points) face landmarks to detect the facial locations such as lips, eyes, nose, etc.

b) Image Preprocessing

Before applying the face landmark detection algorithm, image preprocessing is done. The input RGB image is converted to grayscale, rescaled into 224 × 224 pixels and then converted to torch images. Converting images to grayscale aids in the simplification of algorithms as well as the elimination of complications associated with computational requirements, and it reduces an image to its most basic pixel and thus improves the visualization. Image resizing is done to convert all the images to 224 × 244 pixels and it helps in faster training as deep learning models, in general, train quicker on small images. Here, NumPy images are converted to torch images because for large computations, it speeds up the process. Figure 5.2 shows a sample image from the dataset and the preprocessing step.

c) Face Landmark Detection

Face landmark detection is performed after picture preprocessing. The preprocessed photos, along with the face key points, are sent into the Convolutional Neural Network (CNN) model for this purpose. CNN is a deep neural network that has demonstrated excellent performance in

Figure 5.2 Image preprocessing.

picture categorization, object identification, object localization, and neural style transfer. A convolutional layer, which is the fundamental component of a CNN, enables it to learn spatial information by applying a few filters (or kernels), the parameters of which must be learned during the training process. In this work, a custom CNN model is used since when compared to using pre-trained models and generic architectures, this model with 4 CNN layers provides better performance, greater flexibility, interpretability, and accuracy and adds unique characteristics [22].

Using 4 CNN layers for facial landmark detection is a good decision because it provides for feature extraction, hierarchical representation, regularization, and excellent performance. The appropriate number of layers, however, varies depending on the dataset and task and is frequently established through experimentation and hyperparameter optimization. The CNN architecture consists of 4 convolution layers and 5 fully connected layers. The task of finding facial key points might be difficult because of changes in lighting, face expressions, and positions. Having 4 CNN layers improves the model performance in feature extraction, hierarchical representation, and regularization. A total of 32 4 × 4 filters are included in the first convolutional layer, 64 3 × 3 filters are present in the second, 128 2 × 2 filters are present in the third, and 256 1 × 1 filters are present in the final layer. Stride and padding are both set to 1 for all filters in all layers. Padding is set to 0.

Typically, the filters are smaller than the original image. Each filter engages the image to create an activation map. A convolution layer is used to process the input image in order to extract its features. The activation maps from the final convolutional layer are flattened and transferred to the first fully connected layer. The first layer has 43,264 neurons, whereas the subsequent layers each have 1000 neurons. The final completely connected layer generates 136 points in total, 68 of which are x and 68 of which are y points that correlate to the facial key points. The architecture includes a dropout layer, a 2 × 2 max pooling layer, and a batch normalizing layer in between the convolution layers.

Instead of using raw input as a normalizer, batch normalization is a method utilized between layers of neural networks. Instead of doing everything at once, it is done in little chunks. By accelerating training and leveraging faster learning rates, it facilitates learning. Max pooling is a pooling method that computes the maximum value for each patch in a feature map and then utilizes that value to construct a down sampled feature map. It is often used after a convolutional layer. It adds a little amount of translation invariance, which means that a minor amount of translation has little effect on the values of most pooled outputs. As a mask, the Dropout layer eliminates some neurons' contributions to the following layer while leaving the rest intact. By preventing overfitting on the training data, dropout layers support CNN training. The first set of training data has a disproportionately strong impact on learning if they are absent. Figure 5.3 shows the architectural layout of the CNN model. The output generated by the model displays the 68 facial key points with the points displayed on the input image labeled with numbers from 1 to 68 and are plotted as shown below in Figure 5.4.

d) Masking and Wrapping
Among the facial key points, points 48–60 are the key points for the lips. Thus, using these key points, the lip region is masked as shown in

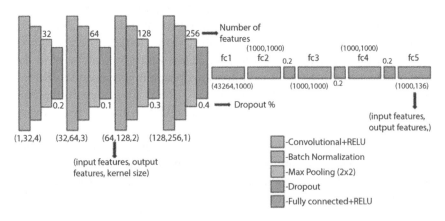

Figure 5.3 CNN architectural diagram.

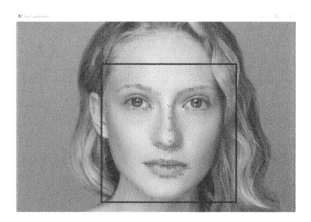

Figure 5.4 Face key points.

Figure 5.5. A color selector bar is used to select the colors as shown in Figure 5.6. A color selector bar is implemented using a trackbar. The base colors used are Red (R), Green (G), and Blue (B). By modifying the different compositions of RGB, we get different colors. After getting the colors, we wrap up the masked region with the color.

This wrapped masked color is placed on the original image after blurring the edges to get a fine finish and produce the final output as shown in Figure 5.7. Similarly, we have implemented the same way for eyebrows and eyeliner. The key points for eyebrows are 17–21 (left eyebrow) and 22–27 (right eyebrow) as in Figure 5.8. The key points for eyeliner are 36 and 39–41 (left eye) and 42 and 45–47 (right eye) as in Figure 5.9.

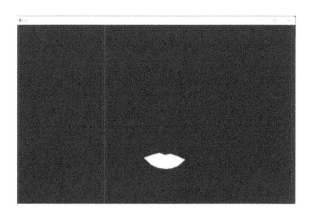

Figure 5.5 Masked image.

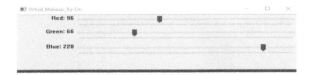

Figure 5.6 Color selector bar.

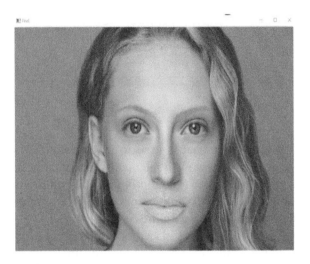

Figure 5.7 Final output image for lipstick.

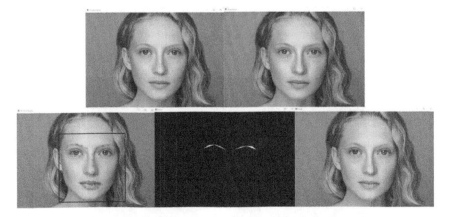

Figure 5.8 Output for eyebrows.

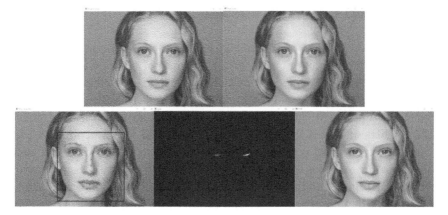

Figure 5.9 Output for eyeliner.

The ability of the model for real-time virtual makeup try-on is tested by capturing the image using a webcam and then applying the face landmark detection, masking, and wrapping steps to get the final output as shown in Figure 5.10.

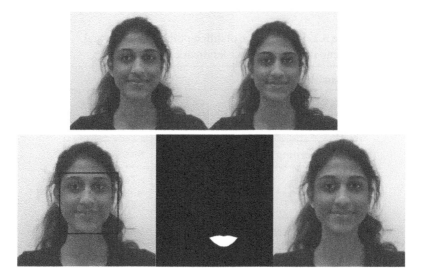

Figure 5.10 Real-time output.

5.4   Experimental Results and Analysis

The proposed model is implemented using an Intel i5 processor - 10210U CPU, which has a frequency of 1.60 GHz, a memory of 8 GB, and an SSD of 477 GB. We implemented the model in Visual Studio using Python version 3.6 and PyTorch is used to build deep neural networks. We utilized the Adam optimizer with a batch size of 10 and an initial learning rate of 0.001. The model is trained for 40 epochs and L1 loss function is used. The suggested model is evaluated in comparison to other traditional and deep learning techniques (like HOG + SVM, Vgg16, Vgg19, Resnet18, and Deep Learning Technique). All the models used in this work are evaluated using the accuracy metric. The proposed model used proves to be more accurate and efficient for virtual makeup try-on using deep learning with an accuracy of 98.5%. The other deep learning models are also trained using the same parameters as the proposed model. The performance of the suggested model and other models are shown in Table 5.2.

Experiments were conducted to find the optimal number of epochs by varying the epochs from 10, 20, 30, and 40. The optimal number of epochs used here is 40 since the graph is more stable after this point. The training and validation accuracy plotted against the number of epochs is shown in Figure 5.11.

Table 5.2 Performance of different models for virtual makeup try-on using face landmark detection.

Algorithm	Accuracy
HOG + SVM [3]	92.68%
Vgg16 [6]	90.84%
Deep Learning Technique (CNN) [7]	89.9%
Resnet18 [1]	87%
Vgg19 [11]	84%
Proposed CNN Model	98.5%

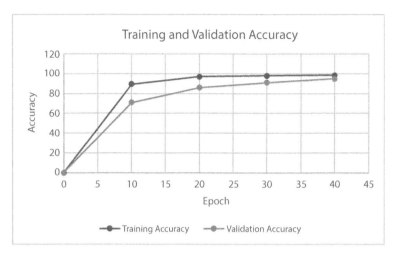

Figure 5.11 Accuracy in training and validation.

5.5   Conclusion

This chapter proposes a virtual makeup try-on system using a deep CNN. The proposed system identifies the facial landmarks, masks the required facial region, and then wraps up the region with the makeup product. The proposed system has been tested using images from a dataset and real-world images. The proposed model exhibited better performance than other deep and image processing-based methods for facial landmark detection. This try-on system can help promote makeup brands and improve sales. Furthermore, it can help users to visualize how makeup looks on them without having to visit the store and physically applying the makeup product. In the future, this model can be expanded to an application or a website that can be used by makeup brands for real-time application of their makeup products.

References

1. Marelli, D., Bianco, S., Ciocca, G., Designing an AI-based virtual try-on web application. *Sensors*, 10, 3832–3858, 2022.
2. Borges, A.D.F.S. and Morimoto, C.H., A virtual makeup augmented reality system. *21st Symposium on Virtual and Augmented Reality (SVR)*, pp. 34–42, 2019.

3. Hung, M.-H., Yang, J., Hsieh, C.-H., A new virtual makeup system based on golden sample search. *Proceedings of the 2020 4th International Conference on Electronic Information Technology and Computer Engineering*, November 2020.

4. Park, J., Kim, H., Ji, S., Hwang, E., An automatic virtual makeup scheme based on personal color analysis. *IMCOM '18: Proceedings of the 12th International Conference on Ubiquitous Information Management and Communication*, pp. 1–7, Article No.: 63, January 2018.

5. Kim, M.C. and Lee, J.-H., Color reproduction in virtual lip makeup using a convolutional neural network. *Color Res. Appl.*, 45, 6, 1190–1201, December 2020.

6. Lin, S., Masharani, N., Renn, M., Shaw, L., Applying virtual makeup using makeup detection and recommendations. *Tech. Disclosure Commons*, 1–10, May 2021, https://www.tdcommons.org/dpubs_series/4276.

7. Ma, X., Zhang, F., Wei, H., Xu, L., Deep learning method for makeup style transfer: A survey. *Cognit. Rob.*, 1, 182–187, 2021.

8. Pang, S., Tao, X., Xiong, N.N., Dong, Y., An efficient style virtual try on network for clothing business industry. 1–10, May 2021, arXiv preprint arXiv:2105.13183.

9. Minar, M.R., Tuan, T.T., Ahn, H., Rosin, P.L., Lai, Y.-K., CP-VTON+: clothing shape and texture preserving image-based virtual try-on. *The IEEE/CVF Conference on Computer Vision and Pattern Recognition (CVPR) Workshops*, June 2020.

10. Neuberger, A., Borenstein, E., Hilleli, B., Oks, E., Alpert, S., Image based virtual try-on network from unpaired data. In *Proceedings of the IEEE/CVF Conference on Computer Vision and Pattern Recognition*, pp. 5184–5193, 2020.

11. Han, X., Wu, Z., Wu, Z., Yu, R., Davis, L.S., VITON: An image-based virtual try-on network. In *Proceedings of the IEEE Conference on Computer Vision and Pattern Recognition*, pp. 7543–7552, 2018.

12. Dong, H. *et al.*, Towards multi-pose guided virtual try-on network. *IEEE/CVF International Conference on Computer Vision (ICCV)*, pp. 9025–9034, 2019, doi: 10.1109/ICCV.2019.00912.

13. Gustavsson, O., Ziegler, T., Welle, M.C., Bütepage, J., Varava, A., Kragic, D., Cloth manipulation based on category classification and landmark detection. *Int. J. Adv. Rob. Syst.*, 19, 4, 1–17, 2022.

14. Z., T., Butepage, J., Welle, M.C., Varava, A., Novkovic, T., Kragic, D., Fashion landmark detection and category classification for robotics. *IEEE International Conference on Autonomous Robot Systems and Competitions (ICARSC)*, pp. 81–88, 2020.

15. Zhu, H., Tong, J., Zhang, L., Zou, X., Research and development of virtual try-on system based on mobile platform. *International Conference on Virtual Reality and Visualization (ICVRV)*, pp. 406–407, October 2017.

16. Roy, D., Santra, S., Mukherjee, D., Chanda, B., Significance of skeleton-based features in virtual try-on. *IEEE Trans. Emerg. Top. Comput. Intell.*, 3, 362–377.

17. Wolf, L., Hassner, T., Maoz, I., Face recognition in unconstrained videos with matched background similarity. *IEEE Conf. on Computer Vision and Pattern Recognition (CVPR)*, pp. 529–534, 2011.

18. Bulat, A. and Tzimiropoulos, G., How far are we from solving the 2D & 3D Face Alignment problem? (and a dataset of 230,000 3D facial landmarks). In *Proceedings of the IEEE International Conference on Computer Vision*, pp. 1021–1030, 2017.

19. Zhao, F., Xie, Z., Kampffmeyer, M., Dong, H., Han, S., Zheng, T., Zhang, T., Liang, X., M3D-VTON: A monocular-to-3D virtual try-on network. *ICCV*, 2021.

20. Hsu, C.-F., Lin, C.-C., Hung, T.-Y., Lei, C.-L., Chen, K.-T., A detailed look at CNN-based approaches in facial landmark detection. 1–9, 2020, arXiv preprint arXiv:2005.08649.

21. Wang, B., Zheng, H., Liang, X., Chen, Y., Lin, L., Yang, M., Toward characteristic-preserving image-based virtual try-on network. *ECCV*, pp. 1–16, 2018.

22. He, S., Song, Y.-Z., Xiang, T., Style-based global appearance flow for virtual try-on. *Conference on Computer Vision and Pattern Recognition (CVPR)*, pp. 483–494, 2022.

6

Reinforcement Learning for Demand Forecasting and Customized Services

Sini Raj Pulari[1], T. S. Murugesh[2]*, Shriram K. Vasudevan[3] and Akshay Bhuvaneswari Ramakrishnan[4]

[1]Bahrain Polytechnic, ISA Town, Bahrain
[2]Department of Electronics and Communication Engineering, Government College of Engineering Srirangam, Tiruchirappalli, Tamil Nadu, India (On Deputation from Annamalai University, Department of Electronics and Instrumentation Engineering, Faculty of Engineering & Technology, India)
[3]Lead Technical – Evangelist (Asia Pacific and Japan), Intel India Pvt. Ltd., Bengaluru, Karnataka, India
[4]Department of Computer Science and Engineering, Sastra Deemed to be University, SASTRA University Thanjavur Campus, Thanjavur, Tamil Nadu, India

Abstract

Reinforcement Learning (RL) is a strong way for machines to learn that has shown promise in areas like predicting demand and providing personalized services. This chapter investigates how strategies based on RL can improve the accuracy of demand forecasting and make it possible for businesses to provide individualized services to each of its clients. The principles of RL, its use in demand forecasting, and the implementation of individualized services are covered extensively as key components of the topic. A real-life case study of a large retail chain highlights the practical benefits of applying RL in optimizing inventory management and providing individualized product recommendations. Researchers at the University of California, Berkeley, carried out this case study. RL gives organizations the ability to dynamically respond to shifting market dynamics and the tastes of their customers by empowering them to continuously learn and adapt to new circumstances. This gives them an advantage over their rivals. This chapter offers light on the revolutionary influence that RL has had and presents a data-driven strategy to meet the demands of modern business environments.

**Corresponding author*: tsmurugesh@gmail.com

Elakkiya, R. and Subramaniyaswamy V. (eds.) Cognitive Analytics and Reinforcement Learning: Theories, Techniques and Applications, (123–134) © 2024 Scrivener Publishing LLC

Keywords: Reinforcement learning, demand forecasting, personalized services, machine learning, inventory management, customer experience, continuous learning, adaptive systems, retail industry

6.1 Introduction

Businesses in today's fast-paced and data-driven world confront ever-increasing hurdles when it comes to effectively predicting customer demand and providing individualized services to their clientele. Traditional methods of forecasting frequently fail to adequately capture the complexities of dynamic markets, and systems that are one size fits all are unable to respond to specific consumers' tastes. However, the advent of Reinforcement Learning (RL) has made it possible to revolutionize demand forecasting and service personalization [1] in exciting new ways. A subfield of machine learning known as reinforcement learning focuses on the process of making successive judgments in an environment with the goal of increasing cumulative rewards. This one-of-a-kind method enables companies to get knowledge from their interactions with the surrounding environment and to continuously adjust their plans to achieve better results. Companies are able to obtain deeper insights into customer behavior, better inventory management, and respond more nimbly to variations in the market when they use RL approaches to the demand forecasting process [2]. In addition to this, RL makes it possible to provide customized services to each individual customer. RL-driven recommendation systems are able to deliver individualized product suggestions to customers by analyzing massive amounts of data and learning from previous encounters. This results in improved customer experiences and the development of consumer loyalty. In this work, we investigate the possibilities that RL present in the areas of demand forecasting and tailored service delivery. In this chapter, we look into the principles of RL, as well as its adaption for demand forecasting and its use in real-life scenarios. We highlight the practical benefits of RL through a captivating case study of a large retail chain. We show how it enables businesses to remain competitive, achieve higher operational efficiency, and please customers in a marketplace that is becoming increasingly dynamic and diversified. Enterprises have the ability to unlock new horizons in data-driven decision-making and alter their strategies in order to survive in the age of personalized services if they embrace the potential of RL.

6.2   RL Fundamentals

The term "reinforcement learning" (RL) refers to an effective machine learning paradigm that is categorized under the more general heading of "artificial intelligence." In contrast to supervised learning, in which models are trained on labeled data, and unsupervised learning, which focuses on detecting patterns in unlabeled data, reinforcement learning (RL) is centered on making sequential decisions to maximize cumulative rewards in an environment [3]. Both types of learning can be divided into three categories: supervised learning, unsupervised learning, and reinforcement learning. The concept of an agent interacting with their surroundings might be considered the driving force behind RL. The agent takes actions based on a policy, makes observations about the present state of the environment, and then receives feedback in the form of rewards or penalties for those behaviors. The agent's objective is to figure out the best possible strategy for maximizing their long-term profits over the course of time. A detailed flowchart is presented in Figure 6.1.

a) Principal Constituents of RL
A representation of the current state or configuration of the environment that the agent is observing is referred to as the state of the environment.

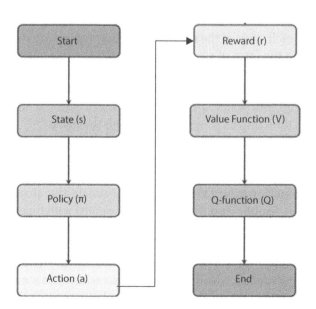

Figure 6.1 Reinforcement learning—key components flowchart.

It is the data that is fed into the RL algorithm, which means that it is a vital source of information for decision-making [4]. An action is a choice that the agent can make in response to a certain situation. Actions are available to the agent. The objective of the agent is to figure out what steps to take in order to get the most desirable results. The agent's method for selecting actions based on the observed states is defined by the policy, which is referred to as "the policy." It leads the agent's decision-making process and maps states into actions at the same time. After carrying out an action, the agent will receive a feedback signal from the surrounding environment, which is referred to as the reward. It expresses the desirability of the agent's conduct in the particular state that is being considered. The value function calculates an estimate of the predicted long-term cumulative reward that an agent can achieve from a given state by adhering to a particular policy. This reward can be thought of as an accumulation of smaller rewards over time. It provides the agent with assistance in evaluating the desirability of various states.

The Q-function calculates the predicted long-term cumulative payoff for carrying out a specific action while in a particular state and adhering to a predetermined policy. It provides the agent with assistance in determining the attractiveness of various action-state combinations. Imagine that you have a self-aware robot that needs to find its way out of a labyrinth to complete a specific mission. The maze setting serves as the RL environment, and the robot functions as the RL agent in this scenario. The robot is provided with observations (state) of its current position and surroundings at each stage of the maze. These observations are represented as a coordinate in the form of x and y. The activities that are open to the robot are moving up, down, left, or right. The action that is selected determines the robot's next state, which moves it to a cell that is adjacent to its current location. The robot's policy specifies the course of action it should take in each state, so directing its progression through the maze.

As the robot navigates the maze, it earns positive incentives when it successfully reaches the goal and receives negative rewards when it comes into contact with obstacles or when it crosses the bounds of the maze. The robot employs a value function (V-function) to estimate the attractiveness of each condition, leading its decisions to choose paths with higher estimated values. This allows the robot to optimize its pathfinding and improve its efficiency. In addition to this, the robot makes use of a Q-function, which allows it to evaluate the worth of particular actions in particular situations. This enables it to decide which action to do at each step in a more accurate manner. The robot is able to identify an optimal way to reach the goal effectively and avoid obstacles during its navigation through the maze because it learns from the rewards it receives and explores a variety of possible routes. This allows the robot to change its policy and value functions over time represented in Figure 6.2.

b) RL Algorithms

There are several different RL algorithms, including model-free and model-based methods of operation. Model-free algorithms, such as Q-learning

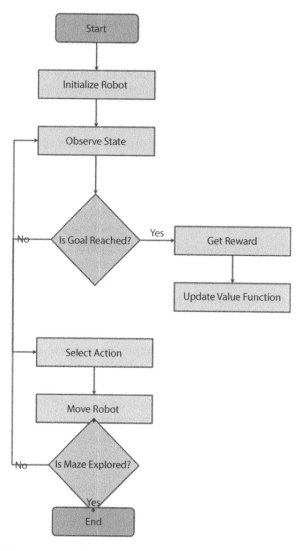

Figure 6.2 Robot working in the maze.

and Deep Q Networks (DQNs), can directly learn the best policy to implement without first having to consciously construct a model of the surrounding environment [5]. On the other hand, model-based algorithms, such as Dynamic Programming and Monte Carlo approaches, work to improve decision-making by constructing a model of the surrounding environment. Available are algorithms that do not require a neutron model. Several well-known algorithms are discussed in this chapter: Q-learning is a well-known off-policy RL technique with the goal of directly learning the optimal action-value function, also known as the Q-function. Estimating the Q-values for each state-action pair is done by a table or a function approximation, respectively. The Q-learning algorithm modifies the Q-values based on the Bellman equation, gradually enhancing the effectiveness of the policy over time. Deep Q-learning Networks (DQNs) take Q-learning to higher dimensional state spaces by approximating the Q-function with the help of deep neural networks.

The input that the neural network receives is the state, and it produces Q-values for each action. The learning process is improved and made more stable by using experience replay and target networks, both of which are employed by DQNs. Policy gradient approaches, such as REINFORCE and Proximal Policy Optimization (PPO), try to learn the policy directly without explicitly calculating value functions. This is their primary objective. They use gradient ascent to optimize the parameters of the policy so that they can maximize the expected cumulative payoff. RL Based on Models Estimating value functions can be done with algorithms such as the Monte Carlo approach by using episodes of contact with the environment. Monte Carlo approaches approximate both state-value (V-function) and action-value (Q-function) functions by taking the average return from a set of sampled trajectories and averaging it. In order to update value functions based on bootstrapping, TD approaches incorporate concepts derived from dynamic programming and methodologies developed by Monte Carlo. Estimating value functions by making use of one-step look ahead updates is what the TD(0) approach is all about. Model Predictive Control (MPC) and similar algorithms simulate trajectories in order to identify the appropriate actions to take based on the learned model.

c) Exploration vs. Exploitation

Finding a happy medium between using existing resources and discovering new ones is one of the most difficult challenges in RL. In contrast, exploitation requires selecting the most well-known activities in order to maximize one's immediate rewards, whereas exploration entails attempting novel acts in order to gather knowledge about the surrounding environment [6].

The ability of an RL agent to strike a balance between these two elements is essential if the agent is to successfully discover optimal tactics while also achieving good performance in the short term. Because it lays the groundwork for the creation of complex RL-based solutions in a variety of fields [7], having a solid understanding of these fundamental concepts is absolutely necessary in order to fully appreciate the potential of RL in demand forecasting and tailored service delivery. The difficulty lies in locating that sweet spot when exploration and exploitation are perfectly complementary to one another. An approach that places an excessive amount of emphasis on exploration may result in an excessive amount of trial and error, which may lead to lost time and performance that is less than optimal. On the other hand, placing an excessive amount of emphasis on exploitation can lead to the agent missing out on the opportunity to uncover potentially better actions and can restrict its capacity to adapt to changing circumstances.

The fact that the exploration–exploitation balance might shift during the course of RL's existence is a key component of the simulation. Agents will frequently begin their time in a new environment by focusing on exploration in order to learn as much as they can about it. As they gain more experience, however, they will progressively shift their focus to exploitation. The best amount of exploration to perform in a given situation could be different based on the circumstances or the current state of the environment. Exploration that is dependent on its context grants the agent the ability to modify its exploration strategy in response to particular circumstances or conditions. The core conundrum of Reinforcement Learning is the debate between exploration and exploitation, which has substantial repercussions for the learning efficiency and overall performance. It is essential for agents to find the appropriate balance in order for them to learn optimal policies and attain long-term rewards in settings that are dynamic and uncertain. RL agents are able to explore and exploit their environments in an effective manner thanks to adaptive and context-dependent exploration techniques. Figure 6.3 explains each stage in the form of a flowchart. These strategies facilitate the finding of optimal solutions and provide the agent the ability to make educated judgments in complicated real-world circumstances.

d) Learning Process
The learning process in Reinforcement Learning (RL) involves initializing the agent's policy and value functions, allowing it to interact with its environment, pick actions based on its policy, get rewards, and then update its value function. Throughout the entirety of the procedure, the agent makes it a point to maintain a healthy equilibrium between exploratory and exploitative behavior, gaining insight not only from the rewards it

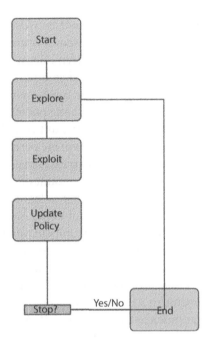

Figure 6.3 Stage of the exploration vs. exploitation process.

obtains but also from the state transitions it experiences. Through a variety of interactions, the agent is able to continuously update its policy and value functions, which brings it closer and closer to an ideal approach for achieving its objectives. The RL agent continuously improves its performance by repeatedly improving its decision-making abilities, adjusting to complex and dynamic situations, and maximizing cumulative rewards. This makes RL a powerful technique for a variety of real-world applications since it continually learns and adapts to its surroundings.

6.3   Demand Forecasting and Customized Services

Reinforcement Learning (RL) is a method that provides a robust and cutting-edge solution to the challenges that modern-day business operations have in accurately predicting customer demand. Traditional methods frequently struggle to keep up with dynamic marketplaces and shifting consumer behavior, which is why RL is such an attractive alternative. RL enables organizations to make data-driven decisions based on historical data and real-time feedback, which ultimately leads to more accurate predictions and increased flexibility. This is accomplished by the modeling of demand forecasting as a sequential decision-making process. When it comes to demand forecasting, some of the most important aspects of RL are the state representation, action space, reward function, policy design, and value function. RL implementation, despite the potential benefits it could provide, comes with a number of problems [8], including the management of exploration–exploitation trade-offs, the processing of large-scale data, and addressing concerns related to data quality. Nonetheless, RL's real-world applications in a variety of industries, like as retail and supply chain management, have proved its capability to optimize inventory management, enhance customer satisfaction, and drive the success of businesses. The impact that RL is having on demand forecasting is poised to change corporate operations and promote efficiency and competitiveness on a worldwide scale. This is because RL is continuing to evolve. Customized services have emerged as a major component of contemporary company strategy, with the overarching objective of catering to the specific tastes of individual customers and improving the quality of the user experience as a whole. Reinforcement learning (RL) is a paradigm that shows promise for the delivery of tailored services that are able to adapt to the specific requirements and behaviors of each individual consumer. With the help of RL, companies are able to analyze enormous amounts of client data, gain insight from previous contacts with customers, and then provide dynamic and customized service recommendations. State representation, action space that defines service possibilities, reward function that is dependent on customer satisfaction, policy design, and value function that evaluates customized methods [9] are the essential components of RL for customized services. It is possible for businesses to give real-time and context-aware product suggestions, personalized content, and adaptive pricing with RL-driven recommendation systems. This eventually leads to greater consumer loyalty and long-term value. However, in order to ensure the ethical and appropriate use of customer data, difficulties such as concerns around privacy and data security need to be carefully addressed. Because of its potential to transform personalized services across industries, RL will continue to be a driving force in the development of customer-centric business strategies in the era of digital technology.

6.4 eMart: Forecasting of a Real-World Scenario

In this analysis of a real-world scenario, the fictitious e-commerce behemoth known as "eMart" implements an effective combination of Demand Forecasting, Natural Language Processing (NLP), and a Recommendation

System in order to improve its company operations and provide more individualized services to its clientele. eMart improves its predictive capabilities by assessing previous sales data, customer behavior, and external factors in order to optimize inventory management and satisfy consumer demand. This is accomplished through the utilization of reinforcement learning for demand forecasting. After that, Natural Language Processing (NLP) will process the unstructured text input and will extract client preferences and feelings in order to develop unique profiles. These profiles, along with the insights gleaned through Demand Forecasting, power a complex RL-driven Recommendation System that provides clients with individualized product recommendations, promotions, and special content that is specifically catered to their needs. Because of the system's ongoing learning and adaptation, customer pleasure, engagement, and loyalty have all grown. The workflow of this company is presented as a flowchart in Figure 6.4. This places eMart at the forefront of personalized e-commerce experiences in an industry that is very competitive.

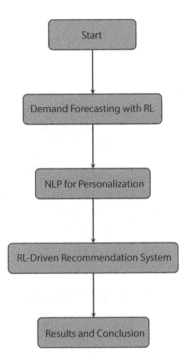

Figure 6.4 Workflow diagram.

6.5   Conclusion and Future Works

This work focuses on the transformative impact that Reinforcement Learning (RL) has had in demand forecasting and tailored services. It offers crucial insights into how organizations can optimize their operations and the experiences they provide for their customers. Demand forecasting becomes more accurate and adaptive because of RL's unique capacity to make sequential decisions and learn from encounters. This enables organizations to adjust dynamically to changing market conditions and client needs. By incorporating RL within the context of personalized services, businesses are able to provide customers with recommendations, promotions, and content that are specific to their needs. This results in increased levels of customer happiness, loyalty, and overall engagement. One of the most important things that you should take away from reading this chapter is the realization that traditional methods of forecasting frequently fall short when it comes to tackling the intricacies of dynamic marketplaces and changing consumer habits. RL provides a data-driven and proactive approach that takes into consideration a variety of elements, including past sales data, external circumstances, and the activities of competitors. This results in enhanced demand predictions and efficient inventory management.

References

1. Wang, P., Chan, C.Y., de La Fortelle, A., A reinforcement learning based approach for automated lane change maneuvers, in: *2018 IEEE Intelligent Vehicles Symposium (IV)*, IEEE, pp. 1379–1384, 2018 Jun 26.
2. Chien, C.F., Lin, Y.S., Lin, S.K., Deep reinforcement learning for selecting demand forecast models to empower Industry 3.5 and an empirical study for a semiconductor component distributor. *Int. J. Prod. Res.*, 58, 9, 2784–804, 2020 May 2.
3. Ding, Z., Huang, Y., Yuan, H., Dong, H., Introduction to reinforcement learning. *Deep Reinforcement Learning: Fundamentals, Research and Applications*, 47–123, 2020. Springer, Singapore. DOI:10.1007/978-981-15-4095-0_2
4. Shin, M., Ryu, K., Jung, M., Reinforcement learning approach to goal-regulation in a self-evolutionary manufacturing system. *Expert Syst. Appl.*, 39, 10, 8736–43, 2012 Aug 1.
5. Oh, J., Hessel, M., Czarnecki, W.M., Xu, Z., van Hasselt, H.P., Singh, S., Silver, D., Discovering reinforcement learning algorithms. *Adv. Neural Inf. Process. Syst.*, 33, 1060–70, 2020.

6. Gupta, A., Mendonca, R., Liu, Y., Abbeel, P., Levine, S., Meta-reinforcement learning of structured exploration strategies. *Adv. Neural Inf. Process. Syst.*, 31, 1–10, 2018.

7. Ishii, S., Yoshida, W., Yoshimoto, J., Control of exploitation–exploration meta-parameter in reinforcement learning. *Neural Networks*, 15, 4-6, 665–87, 2002 Jun 1.

8. Chien, C.F., Lin, Y.S., Lin, S.K., Deep reinforcement learning for selecting demand forecast models to empower Industry 3.5 and an empirical study for a semiconductor component distributor. *Int. J. Prod. Res.*, 58, 9, 2784–804, 2020 May 2.

9. Huang, W., Li, S., Wang, S., Li, H., An improved adaptive service function chain mapping method based on deep reinforcement learning. *Electronics*, 12, 6, 1307, 2023 Mar 9.

7

COVID-19 Detection through CT Scan Image Analysis: A Transfer Learning Approach with Ensemble Technique

P. Padmakumari*, S. Vidivelli and P. Shanthi

School of Computing, SASTRA University, Thanjavur, Tamil Nadu, India

Abstract

The COVID-19 pandemic has had a profound impact on the world due to the highly contagious nature of the virus. To effectively control its spread, identifying and treating patients in the early stages of the disease is crucial. Although reverse transcription-polymerase chain reaction (RT-PCR) is the standard method for COVID-19 detection, it can be prone to errors and time-consuming. To address this issue, deep learning techniques and computed tomography (CT) scans are being used for automated COVID-19 detection. This method involves using the Transfer Learning technique VGG-16 to extract essential features from 2482 CT scan images. Prior to extraction, the images are enhanced through Contrast-limited adaptive histogram equalization (CLAHE). The extracted features are then utilized to train various Machine Learning (ML) models, including Support Vector Machine, Logistic Regression, and Gradient Boosting models. To further improve performance, the identification results from each model are combined in an ensemble learning model using the Hard voting ensemble. This approach achieved an accuracy rate of 98.927%.

Keywords: COVID-19, CLAHE, ResNet, DenseNet, VGG-16, inception, feature extraction, SVM

Corresponding author: padmalec.sastra@gmail.com

Elakkiya, R. and Subramaniyaswamy V. (eds.) Cognitive Analytics and Reinforcement Learning: Theories, Techniques and Applications, (135–150) © 2024 Scrivener Publishing LLC

7.1 Introduction

At the end of December 2019, Wuhan, China made the announcement that they have identified their first instance of a new coronavirus infection (COVID-19). COVID-19 quickly become a widespread problem over the whole planet. SARS-CoV-2 has been spread by bats, causing major acute respiratory problems. Flu, shortness of breath, dry cough, headaches, loss of smell, and taste are among the most prevalent COVID-19 symptoms. At the beginning of the diagnosis process, medical professionals will analyze the patient's symptoms to assess whether or not the patient is affected by COVID-19. Because of this, individuals are at a high risk of being infected with the coronavirus if they come into contact with a droplet. Isolating a person who has a virus and preventing them from coming into contact with healthy individuals is the only method to stop the virus from spreading to healthy people. In the beginning, the Real-Time Reverse Transcription Polymerase Chain Reaction also known as RT-PCR was the one and only method for identifying COVID-19 using respiratory specimens. Because the findings of laboratory tests are reliable, they take a very long time to give an update, and it is not possible to carry out mass screening at a pace that is quicker than the rate at which the virus is spreading. People who exhibit some of these clinical symptoms are subjected to physical examinations and diagnostic procedures such as CT scans and X-rays of the chest. In the end, physicians or medical specialists review COVID-19 results from various tests.

At this point in time, CT scans are more accurate and cost around five times less than PCR testing, which makes them an appealing alternative. It might be challenging to correctly interpret medical images due to the fact that the disease is still in its infancy and has many symptoms with other respiratory conditions, such as pneumonia. To accurately diagnose COVID-19, radiologists must have extensive training and experience in order to perform at an appropriate standard of diagnostic accuracy. In order to manually diagnose COVID-19, the process is not only time-consuming but also prone to errors caused by humans and needs for the assistance of a skilled radiologist. Because the respiratory abnormalities that are associated with COVID-19 may be similar to those that are associated with SARS or Viral Pneumonia (VP), which might delay the detection and treatment of COVID. Because CXR and CT are suggested for a variety of pulmonary problems, any automated solution developed to identify COVID-19 should also take into account other respiratory illnesses in order to produce a more complete and robust diagnostic system [17].

In recent years, there has been a rise in the prevalence of the use of deep learning (DL) strategies in the autonomous processing of radiological images. Deep Learning is a sort of machine learning motivated by the structure of the brain, which is referred to as an Neural Network in Deep Learning. It derives characteristics and tasks from data. Because the aim is to acquire directly from images, it is an end-to-end learning. Traditional neural networks include 2 or 3 hidden layers; however, the most recent networks have up to 150 layers. The fundamental objective of this research is to develop reliable measures for the identification of patients affected by COVID-19 based on CT images processed by CNN. CNN performed quite well in most of the computer vision tasks that it was given. Because of its inherent capacity for self-learning, CNN has recently established itself as the preeminent method of in-depth learning for the classification of medical images. In the course of the study that was carried out, a deep learning base model as well as a conventional machine learning meta model were used. The goal was to enhance classification performance. The meta model made use of the predicted data from the base model as one of its input attributes. Traditional machine learning approaches, such as support vector machine (SVM) and different pre-trained convolutional neural networks (CNN), were used in this experiment as the base model and as the meta model, respectively.

The following are the primary aims of this research:

- CLAHE was used to improve the picture quality of the CT scans.
- Developed a few pre-trained DL models for extracting the most significant characteristics from CT scan images.
- Improved classification performance in terms of accuracy, recall, precision, and AUC by using a Support Vector Machine (SVM) model.

This chapter continues as follows. Section 7.2 describes related work. Section 7.3 presents the methodology and relational algorithm. Section 7.4 covers datasets, experimental conditions, and outcomes. Section 7.5 contains the conclusion remarks.

7.2   Literature Survey

In an effort to speed up diagnosis of this condition, researchers have created automated Deep Learning-based techniques. Deep learning methods are discussed in this section for identifying COVID-19 in X-rays and CT images. Yu *et al.* [1] proposed four pre-trained deep learning models (ResNet-50, ResNet-101, Inception-V3, and DenseNet-201) with numerous classifiers that were utilized to differentiate between severe and non-severe COVID-19 cases. Their classification accuracy was 95.20% and 95.34% for the 10-fold cross-validation and leave-one-out cross-validation procedures. Mishra *et al.* [2] recommended combining the predictions of deep CNN models. The recommended method attained an accuracy of 86% using the COVID-CT dataset (347 COVID-19 against 397 Non-COVID-19). On the other hand, only axial slices were used in this study. Chen *et al.* [3] using radiological semantics and clinical features developed the COVID-19 Diagnostic Models (CDM), which does not require the use of such a nucleic acid test. CDM was shown to be effective in detecting COVID-19 cases, with CDM delivering better diagnostic performance and greater net benefits, as demonstrated in the research. Narin *et al.* [4] used ResNet50, InceptionV3, and Inception ResNetV2 to analyze 50 COVID-19 chest X-ray images and 50 non-invasive images. An accuracy of 98% was attained using the ResNet50 model. The non-COVID photographs in this experiment are those of individuals between the ages of 1 and 5 years old. The Corona virus Recognition Network (CVR-Net) was suggested by Hasan *et al.* [5] as a CNN-based network for detecting various COVID-19 cases from radiology pictures. It was trained and evaluated on X-ray and CT images with such an average accuracy rate of 78% in the CT image dataset. COVID-Net was further tweaked to increase its representational capability and computational efficiency, as said in [6]. Wang *et al.* [7] suggested a novel collaborative learning approach for reliable COVID-19 identification. They upgraded COVID-Net to increase accuracy and learning speed. They also normalized features in latent space. It outperformed COVID-Net. This shows that chest CT and X-rays can detect COVID-19. Existing models employ CNNs because they are good in computer vision and biological imaging.

Yadav *et al.* [8] used data augmentation approaches to assess two pre-trained CNN models, namely, VGG16 and InceptionV3. For binary classifications, the InceptionV3 model scored a classification accuracy of 0.9935, whereas the VGG16 model achieved a classification accuracy of 0.9884. The research [9] recommended DenseNet-OTLS for detecting COVID-19. All three DenseNet121, DenseNet169, and DenseNet201 were employed for the experiment, and 640 photos were used to train and test the model. The author focuses on single and Composite learning rates to enhance the DenseNet model. This model achieved an accuracy of 96.30%. Wang *et al.* [10] proposes a 2D Convolutional model to distinguish COVID-19 and viral pneumonia using manually specified area patches. Ninety-nine patients' chest CT images (44 COVID-19, 55 viral pneumonia) are examined. The testing dataset demonstrates 73.1% accuracy, 67% specificity, and 74% sensitivity. A neural network (COVNet) created by Li *et al.* (2020) [11] was used to identify coronavirus infection in patients by extracting characteristics from chest CT scans. A total of 4357 chest CT scans of 3322 individuals were used to train COVNet. A 95% accuracy rate was achieved from this method. KarNet is a model that was proposed by Arpita Halder and her colleagues [12]. Its backbone is composed of DenseNet201, MobileNet, ResNet50, and VGG16, and it also has additional layers that are used to investigate the performance of each framework independently on origin unaugmented and augmented datasets [13]. They discovered that, out of the four pre-trained models that were utilized in KarNet, the one that makes use of DenseNet201 outperformed the others with a rate of accuracy that was 97%. Table 7.1 tabulates the literature review.

Table 7.1 Literature review.

Study	Models USED	Dataset	Classification accuracy
Yu *et al.* [1]	ResNet-50, ResNet-101, Inception-V3, DenseNet-201	Severe and non-severe COVID-19 cases	95.20% (10-fold cross-validation), 95.34% (leave-one-out cross-validation)
Mishra *et al.* [2]	Deep CNN models	COVID-CT dataset (347 COVID-19 vs. 397 Non-COVID-19)	86%
Chen *et al.* [3]	COVID-19 Diagnostic Models (CDM)	Radiological semantics and clinical features	Better diagnostic performance and greater net benefits
Narin *et al.* [4]	ResNet50, InceptionV3, Inception ResNetV2	50 COVID-19 chest X-ray images and 50 non-invasive images	98% (using ResNet50)

(Continued)

Table 7.1 Literature review. (*Continued*)

Study	Models USED	Dataset	Classification accuracy
Hasan *et al.* [5]	Coronavirus Recognition Network (CVR-Net)	X-ray and CT images	Average accuracy rate of 78% (in CT image dataset)
Wang *et al.* [6]	COVID-Net (upgraded)	Chest CT and X-rays	Improved accuracy and learning speed
Yadav *et al.* [8]	VGG16, InceptionV3	Data augmentation approaches	InceptionV3: 0.9935 accuracy, VGG16: 0.9884 accuracy
Research [9]	DenseNet-OTLS	DenseNet121, DenseNet169, DenseNet201	96.30% accuracy
Wang *et al.* [10]	2D Convolutional model	Chest CT images (44 COVID-19, 55 viral pneumonia)	73.1% accuracy, 67% specificity, 74% sensitivity
Li *et al.* [11]	COVNet	Chest CT scans	95% accuracy
Halder *et al.* [12]	DenseNet201, MobileNet, ResNet50, VGG16 (in KarNet)	Unaugmented and augmented datasets	97% accuracy (DenseNet201)

7.3 Methodology

A deep learning-based decision assistance tool should be developed to rapidly detect COVID-19 patients. A dataset of CT scans from COVID-19 victims has been collected. Because the pictures were not clear, there was a need to employ a variety of methods to pre-process the data. The Contrast Limited Adaptive Image Histogram is a pre-processing method used to enhance the contrast of medical images. After the photos have been pre-processed, the different transfer learning techniques is used to extract

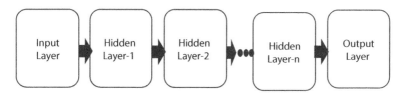

Figure 7.1 Layered architecture of deep learning model.

the most critical features of the image. Finally, a Support Vector Machine is employed to predict the COVID and non-COVID outcomes. Figure 7.1 shows layered architecture of the deep learning model. Sample images of COVID and non-COVID CT scan are shown in Figure 7.2.

a) Pre-processing

Image pre-processing improves the outcome. There are several ways to improve medical imaging. CLAHE was created with the purpose of improving the pictures of medical scans that have poor contrast. The clipping value controls the amount of histogram noise that is smoothed out as well as the amount of contrast that is added. The first thing to do is change the RGB image into a LAB format. A CLAHE approach to the L channel was then used, after which an improved LAB picture was created by merging the enhanced L channel with the A and B channels. In the end, the improved LAB picture was transformed back to the improved RGB image. The figure shows the original CT scan image and the enhanced image after using CLAHE.

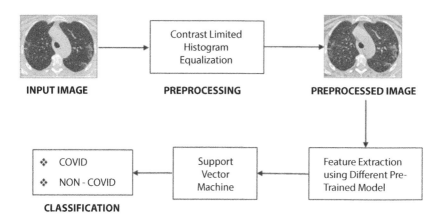

Figure 7.2 Flowchart for classification of COVID-19.

b) VGG-16

Oxford University proposed the VGG-16 deep neural network in 2014 [1]. Convolutional neural networks were relatively new at the time, but this model quickly reached its peak due to its simplicity and usefulness when it was first introduced. After the initial stack of 2 convolution layers with a 3 × 3 receptive size, ReLU activations are performed. The two layers have 64 filters each. One pixel is the convolution stride and padding. Activation maps are then spatial max pooled across a 2 × 2-pixel frame with a 2-pixel stride. It halves activations. The activations run via a second stack with 128 filters vs. 64 in the first. A third stack with three convolutional layers and a max pool layer follows. A total of 256 filters are used. Next are two stacks of three convolutional layers, each with 512 filters. It is a common deep convolutional neural network (CNN) design with numerous layers known as VGG (Visual Geometry Group). VGG-16 or VGG-19 has 16 or 19 convolutional layers, respectively, and is referred to as "deep" in this context.

c) Resnet

The residual network is a deep convolutional network proposed by He Kaiming [9]. It is easier to optimize and can improve the accuracy by increasing the depth of the network. Its core is to solve the problem of gradient degradation caused by an increase in network depth. This can improve network performance by simply increasing the depth of the network. The accuracy of the model may be improved by using residual blocks in deep residual nets. Residual blocks are strong because of the idea of "skip connections," which is at the heart of residual blocks. The Residual Block is shown in Figure 7.3.

d) DenseNet

The acronym "DenseNet" is used to indicate the densely connected convolutional networks. At the CVPR Conference in 2017, Gao Huang and Zhuang Liu, along with the rest of their team, presented their proposal. For every tier, the feature map of all previous layers is employed as inputs, and its own feature map is used as inputs to all following levels; therefore, the first layer is connected to the second, third, fourth, and so on, while the second layer is connected to the third, fourth, fifth, and so on. This is done in order to get the most of the information flow that occurs across the different levels of the network. Rather than only "N" connections, like in standard deep learning designs, the network now has a total of "(N(N+1))/2" connections. DenseNet is built by interconnecting the Dense Block, the convolutional layer, and the pooling layer, with every Dense Block holding 1 * 1 and 3 * 3 convolution kernels.

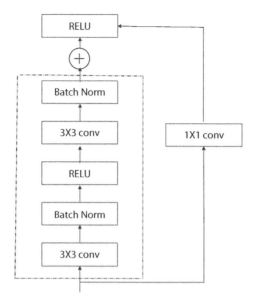

Figure 7.3 Resnet architecture.

e) Inception

GoogLeNet, commonly known as inception, was the winning contribution in the ImageNet 2014 competition, while VGG-16 was the runner-up. Despite the fact that there are 22 levels in total, there is no fully connected layer. The original parameter set was a significant step up from AlexNet's, which had 60 million parameters to choose from, since it represented a significant advance. However, while having a smaller number of parameters, this network has a higher cost in terms of computation. Figure 7.3 depicts the overall layout of the Inception module. Variations of 1 * 1, 3 * 3, and 5 * 5 convolutions may be found. In addition to it, a 3 * 3 pooling layer is also included. With the aid of these Inception modules, the model saves a lot of parameters and reduces overfitting. Figure 7.4 shows inception architecture.

f) Support Vector Machine

Using the Support Vector Machine (SVM) methodology, it is conceivable to construct models for classification as well as regression. On the basis of statistical learning theory, an innovative statistical approach was suggested. The primary goal in data classifications is by using a support vector machine (SVM) to determine a distinctive hyperplane. Using a separating hyperplane, it is possible to split data from a sample into numerous classes

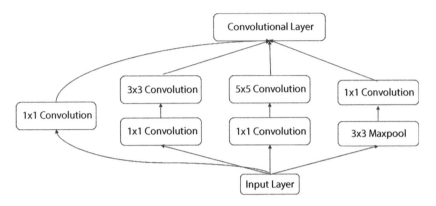

Figure 7.4 Inception architecture.

and ensure that the outcomes of various classes are categorized appropriately. For the purposes of this approach, the support vectors are a subset of the training dataset. Both linear and non-linear data classification may be accomplished using SVMs. SVM classifiers classify non-linear data using kernel functions, which seek to shift the data into a new degree to better categorize it. Many different kernel functions are used by SVM classifiers, including polynomial kernel, RBF kernel, sigmoid kernel, and so on.

7.4   Results and Discussion

The dataset used in this study was obtained by Angelov and Almeida Soares (2020) from hospitals in São Paulo, Brazil. It was made available through the Kaggle competition organized by PlamenEduardo (2020). The dataset consists of a total of 2482 CT scan images, comprising 1252 images from COVID-19-infected patients and 1230 images from non-infected patients. It is important to note that while this dataset includes COVID-19-infected patients, SARS-CoV-2 is not the only cause of pulmonary illness in individuals who do not have COVID-19. Therefore, the dataset also contains images from patients who were scanned for non-COVID-19 indicators. The dataset includes 60 patients who were scanned for COVID-19 infection. Among these patients, there were 32 males and 28 females. At the same time, another group of 60 cases, consisting of 30 men and 30 women, underwent CT scans to check for non-COVID-19 indicators. This gender breakdown suggests an effort to include diversity in the dataset. The dataset provides examples of the data, which may include CT scan images showcasing the characteristics of COVID-19-infected and non-infected patients.

The number of COVID-19-positive and -negative cases in the dataset may consist of 1252 of CT scan images from COVID-19-positive cases and 1230 of images from COVID-19-negative cases. The data collection protocols typically involve acquiring CT scans from patients who have been tested for COVID-19. The protocols may vary across different hospitals and centers. They may involve obtaining informed consent from patients, following specific scanning techniques, and adhering to guidelines provided by medical and imaging professionals. The dataset may include demographic information about the patients, such as age, gender, and potentially other relevant attributes. This information is important for analyzing the impact of COVID-19 across different demographic groups. Figure 7.5 shows the sample images of infected and non-infected patients.

The proposed method was evaluated using a dataset of 2482 CT scan images, consisting of 1241 COVID-19-positive cases and 1241 non-COVID-19 cases. The dataset was split into a training set (70%) and a testing set (30%). The deep learning models (VGG-16, ResNet, DenseNet, Inception) were used to extract features from the CT scan images, and these features were then fed into various machine learning models (Support Vector Machine, Logistic Regression, and Gradient Boosting) for classification. The performance of each individual model was evaluated in terms of accuracy, recall, precision, and area under the ROC curve (AUC). The results obtained for each model are shown in Table 7.2.

Among the individual models, ResNet achieved the highest accuracy, recall, precision, and AUC. However, to further improve performance, an ensemble learning approach using the Hard voting ensemble was employed. The identification results from each individual model were combined, and the final ensemble model achieved an accuracy rate of 98.927%. In this study, a deep learning-based approach using CT scan images and

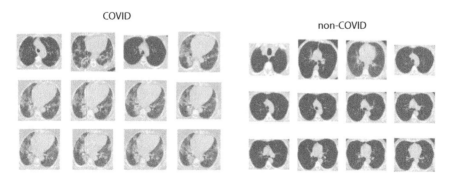

Figure 7.5 Sample images of infected and non-infected patients.

Table 7.2 Result obtained from different models.

Model	Accuracy	Recall	Precision	AUC
VGG-16	95.43%	95.11%	95.76%	0.9703
ResNet	97.15%	96.81%	97.58%	0.9847
DenseNet	96.82%	96.41%	97.22%	0.9836
Inception	96.62%	96.12%	97.16%	0.982

machine learning models was proposed for automated COVID-19 detection. The images were pre-processed using CLAHE to enhance the contrast and improve the quality of the images. Various deep learning models (VGG-16, ResNet, DenseNet, and Inception) were used to extract features from the images, and machine learning models (Support Vector Machine, Logistic Regression, and Gradient Boosting) were employed for classification. The results demonstrated high accuracy rates for the individual models, with ResNet achieving the highest performance. The ensemble learning model further improved the accuracy rate to 98.927%. This approach has the potential to assist medical professionals in the early and accurate detection of COVID-19, especially in situations where PCR testing is time-consuming or not readily available. However, further research and validation on larger and diverse datasets are necessary to validate the effectiveness and generalizability of the proposed method.

With ongoing advancements in deep learning and medical imaging technologies, automated COVID-19 detection systems can play a significant role in combating the pandemic and saving lives. The proposed approach for automated COVID-19 detection using deep learning techniques and CT scans demonstrates several advantages and limitations compared to other methods. The contextual understanding of the effectiveness and uniqueness of the proposed approach has the following advantages and limitations: Improved Efficiency: Compared to the standard method of COVID-19 detection using RT-PCR, which can be time-consuming, the proposed approach offers the advantage of automated detection using CT scans. This can potentially expedite the identification and treatment of COVID-19 patients in the early stages of the disease. Deep Learning Techniques: By employing deep learning techniques, specifically the Transfer Learning technique with VGG-16, the proposed approach leverages pre-trained models to extract essential features from CT scan images.

This allows for efficient and effective feature extraction, which can aid in accurate COVID-19 detection.

a) Enhanced Image Pre-processing

The use of Contrast-limited adaptive histogram equalization (CLAHE) as an image enhancement technique prior to feature extraction helps to improve the quality and visibility of relevant features in CT scan images. This enhancement can contribute to the accuracy of the subsequent machine learning models. Ensemble Learning: The combination of identification results from multiple machine learning models, such as Support Vector Machine, Logistic Regression, and Gradient Boosting, using the Hard voting ensemble technique further enhances the performance of the proposed approach. Ensemble learning helps to reduce biases and errors associated with individual models and can potentially improve overall accuracy.

b) Accuracy

The proposed approach achieved an impressive accuracy rate of 98.927%. This indicates that the approach has the potential to accurately identify COVID-19 cases, which is crucial for effective disease control and management.

c) Generalizability

The effectiveness of the proposed approach may be limited to the specific dataset and context used in the study. It is important to assess the generalizability of the approach by testing it on diverse datasets from different populations and imaging protocols to ensure its effectiveness in real-world scenarios.

d) Interpretability

Deep learning models, such as VGG-16, often lack interpretability, making it challenging to understand the underlying factors contributing to the model's decisions. Interpretability is essential for gaining trust and confidence in the automated detection system, especially in critical medical applications.

e) Dataset Limitations

The discussion does not provide detailed information about the dataset used for training and testing the proposed approach. It is important to consider factors such as dataset size, diversity, and potential biases to ensure the reliability and generalizability of the approach.

f) Sensitivity to Image Quality

The proposed approach heavily relies on the quality of CT scan images, and any noise or artifacts in the images may affect the accuracy of the automated detection. Robustness to variations in image quality, such as different CT machines or imaging protocols, should be carefully evaluated. These limitations can be overcome in future works.

7.5 Conclusion

Patients with COVID-19 and those without it may be identified from CT scan images using binary classification in this study. Medical image analysis relies heavily on computer-aided methods for automatically classifying CT scans. CT image analysis at the microscopic level is difficult and time-consuming. In this study, identifying COVID-19 patients using machine learning-based CT image classification exceeded expectations. At first, CLAHE was used as a pre-processing model to enhance the quality of image. Then, several deep learning models are used as a feature extraction. Once extracted, support vector machine is used for classification. Per the results, the VGG-16 model gives higher accuracy than others. Both accuracy and recall are quite high at 98.93% and 99%, respectively.

The VGG-16-based CNN plus SVM outperforms some of the most well-known deep transfer learning models in comparison. The difference in value of every model may not be very high but a 1% performance gain may save a lot of lives when applied to a larger population. In comparison to standard RT-PCR testing, the COVID-19 detection utilizing this model is substantially more rapid and accurate. It is clear from this finding that the model has delivered on its promise of assisting clinicians in the quick diagnosis of COVID-19 infection and prompt treatment of patients. The key disadvantage of this study is that the model was trained on a small dataset since the COVID-19 patient dataset is restricted. An effort should be made to acquire a huge dataset in the future. In the future, the extracted features are employed by different machine learning algorithms and the classification is done through an ensemble method that could be implemented and tested for further development and assessment for more accurate identification of COVID-19 patients using CT images.

References

1. Yu, Z., Li, X., Sun, H., Wang, J., Zhao, T., Chen, H., Ma, Y., Zhu, S., Xie, Z., Rapid identification of COVID-19 severity in CT scans through classification of deep features. *Biomed. Eng. Online, 19*, 1, 1–13, 2020.
2. Mishra, A.K., Das, S.K., Roy, P., Bandyopadhyay, S., Identifying COVID19 from chest CT images: A deep convolutional neural networks based approach. *J. Healthc. Eng.*, 2020, 1–7, 2020.
3. Chen, H., Guo, J., Wang, C., Luo, F., Yu, X., Zhang, W., Li, J., Zhao, D., Xu, D., Gong, Q., Liao, J., Clinical characteristics and intrauterine vertical transmission potential of COVID-19 infection in nine pregnant women: A retrospective review of medical records. *Lancet, 395*, 10226, 809–815, 2020.
4. Narin, A., Kaya, C., Pamuk, Z., Automatic detection of coronavirus disease (covid-19) using x-ray images and deep convolutional neural networks. *Pattern Anal. Appl., 24*, 1207–1220, 2021.
5. Hasan, M.K., Alam, M.A., Elahi, M.T., Roy, S., Wahid, S.R., CVR- Net: A deep convolutional neural network for coronavirus recognition from chest radiography images. *Electric. Eng. Syst. Sci.*, 11993, 1–31, 2020.
6. Wang, Z., Liu, Q., Dou, Q., Contrastive cross-site learning with redesigned net for COVID-19 CT classification. *IEEE J. Biomed. Health Inf., 24*, 10, 2806–2813, 2020.
7. Yadav, S.S., Bendre, M.R., Vikhe, P.S., Jadhav, S.M., Analysis of deep machine learning algorithms in covid-19 disease diagnosis, 12, 1–16, August 2020.
8. Zhang, Y.D., Satapathy, S.C., Zhang, X., Wang, S.H., COVID-19 diagnosis via DenseNet and optimization of transfer learning setting. *Cogn. Comput.*, 8, 1–17, 2021.
9. Wang, S., Kang, B., Ma, J., Zeng, X., Xiao, M., Guo, J., Cai, M., Yang, J., Li, Y., Meng, X., Xu, B., A deep learning algorithm using CT images to screen for Corona Virus Disease (COVID-19). *Eur. Radiol., 31*, 6096–6104, 2021.
10. Li, L., Qin, L., Xu, Z., Yin, Y., Wang, X., Kong, B., Bai, J., Lu, Y., Fang, Z., Song, Q., Cao, K., Artificial intelligence distinguishes COVID-19 from community acquired pneumonia on chest CT. *Radiology*, 200905, 16, 2020.
11. Halder, A. and Datta, B., COVID-19 detection from lung CT-scan images using transfer learning approach. *Mach. Learn.: Sci. Technol., 2*, 4, 045013, 2021.
12. Atek, S., Pesaresi, C., Eugeni, M., De Vito, C., Cardinale, V., Mecella, M., Rescio, A., Petronzio, L., Vincenzi, A., Pistillo, P., Bianchini, F., A geospatial artificial intelligence and satellite-based earth observation cognitive system in response to COVID-19. *Acta Astronaut., 197*, 323–335, 2022.
13. Amin, J., Sharif, M., Gul, N., Kadry, S., Chakraborty, C., Quantum machine learning architecture for COVID-19 classification based on synthetic data generation using conditional adversarial neural network. *Cogn. Comput., 14*, 5, 1677–1688, 2022.

14. Alshmrani, G.M.M., Ni, Q., Jiang, R., Pervaiz, H., Elshennawy, N.M., A deep learning architecture for multi-class lung diseases classification using chest X-ray (CXR) images. *Alexandria Eng. J.*, *64*, 923–935, 2023.

15. Elakkiya, R., Subramaniyaswamy, V., Vijayakumar, V., Mahanti, A., Cervical cancer diagnostics healthcare system using hybrid object detection adversarial networks. *IEEE J. Biomed. Health Inf.*, 26, 4, 1464–1471, April 2022.

16. Priyadarsini, M.J.P., Rajini, G.K., Hariharan, K., U. Raj, K., Ram, K.B., Indragandhi, V., Subramaniyaswamy, V., Pandya, S., Lung diseases detection using various deep learning algorithms. *J. Healthc. Eng.*, 2023, 1–13, 2023.

17. Elakkiya, R., DK, Jain, Kotecha, K., Pandya, S., SS, Reddy, Varadarajan, V., Mahanti, A., Hybrid deep neural network for handling data imbalance in precursor MicroRNA. *Front. Public Health*, 9, 821410, 2021 Dec 23.

8

Paddy Leaf Classification Using Computational Intelligence

S. Vidivelli*, P. Padmakumari and P. Shanthi

School of Computing, SASTRA University, Thanjavur, Tamil Nadu, India

Abstract

Plant diseases have been a significant problem for farmers, causing severe damage to crops and resulting in economic losses. The detection and classification of plant diseases are essential for maintaining a healthy crop yield and reducing the use of pesticides, which can harm the environment and human health. In this chapter, we propose a framework for the classification of rice leaf diseases using our proposed Local Binary Pattern and Fractal features, extracted from digital images of diseased leaves. To achieve the goal of efficient and accurate disease detection, we utilize modern technologies such as image processing and machine learning techniques. Our proposed framework uses an Adaboost ensemble classifier for classification, which has shown to be effective in several applications. The framework is tested on a publicly available dataset of rice leaf images, and the results show that our proposed method outperforms other state-of-the-art methods in terms of accuracy.

Keywords: Fractal dimension, paddy, Adaboost, adaptive thresholding, support vector machine

8.1 Introduction

Agriculture is a critical driving force for many countries' economies, and its success is largely dependent on environmental factors such as weather and the quality of crops. In modern agriculture, the use of technologies like mechanical devices, moisture sensors, and image processing has

---

Corresponding author: vidivelli@cse.sastra.edu

Elakkiya, R. and Subramaniyaswamy V. (eds.) Cognitive Analytics and Reinforcement Learning: Theories, Techniques and Applications, (151–166) © 2024 Scrivener Publishing LLC

allowed farmers to be more profitable and environmentally friendly. However, in less-developed countries, crop losses from plant diseases can lead to hunger and starvation. In India, agriculture is a major contributor to the country's economy, providing employment to 60% of the population and generating 17% of India's total GDP. Paddy is one of the most important food crops in India and the world, and it is vulnerable to various diseases caused by fungi, bacteria, and viruses [1]. Every year, farmers lose an estimated 37% of their paddy crop to diseases, often because they are unable to accurately identify the disease and choose the appropriate control methods. To improve crop yield and address this problem, an effective and cost-efficient method for disease detection is needed. This study proposes a machine learning approach using features extracted from plant leaves to accurately detect and classify two common rice leaf diseases. This method is less expensive and provides quick and accurate results, enabling farmers to choose the best remedial measures, such as pesticides, to control the disease and improve crop production. Paddy crops are vulnerable to various diseases such as Rice Blast and Bacterial Blight, which can significantly reduce the production of paddy crops. To combat this, it is crucial to have an effective and efficient method for disease detection and classification. In this study, we aim to detect and classify paddy leaf diseases such as Brown Spot, Bacterial Blight, and False Smut as shown in Figure 8.1 [2].

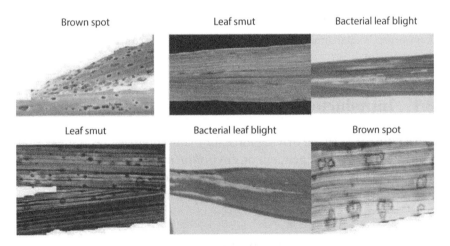

Figure 8.1 Types of diseases.

a) Brown Spot

Brown Spot is caused by the organism *Helminthosporium oryzae* and attacks the rice crop from the seedling stage in the nursery to the milk stage in the field. The symptoms of this disease are circular to oval-shaped dark brown to reddish-brown spots that vary in size and shape and appear on the coleoptiles, leaf sheath, and glumes.

b) Bacterial Blight

Bacterial Blight is caused by the organism *Xanthomonas oryzae* and is one of the most destructive diseases of rice cultivation. It leads to crop loss up to 75% in severe cases. The symptoms of Bacterial Blight include yellowing and wilting of infected leaves, light green to grayish green color lines on leaves, milky bacteria ooze dripping from leaves, and spread through wind and rain splash.

c) False Smut

False Smut is caused by the organism *Ustilaginoidea virens* and is a major grain disease in rice production. This pathogen attacks spikelets and converts them into False Smut balls made up of chlamydospores covered with mycelia. The symptoms of False Smut include yellowish to orange small balls on the grains that later turn black, growth of velvety fragments enclosing floral parts, and damage to the membrane caused by spore growth.

Artificial intelligence (AI) and machine learning (ML) play an increasingly important role in a wide range of industries, including agriculture. In the context of paddy leaf disease detection, AI and ML can be used to develop accurate and efficient detection models. These models can be trained on large datasets of labeled images to recognize patterns and classify images into healthy or diseased categories [3]. The use of AI and ML can significantly improve the accuracy and speed of detection, enabling early intervention and prevention of crop losses. Moreover, these technologies can provide decision support to farmers and experts, allowing them to optimize treatment and management of crops. Overall, AI and ML are powerful tools that can transform agriculture and improve food security around the world.

8.2   Literature Review

Feature extraction is a crucial step in the automatic diagnosis of rice leaf diseases. Several techniques have been proposed for feature extraction, including Local Binary Patterns (LBP), Fractal Dimension (FD), and Lacunarity [4].

These techniques have been applied to rice leaf disease prediction and compared with various classifiers. LBP has been widely used for feature extraction in the field of image analysis, including the diagnosis of rice leaf diseases. LBP describes the texture of an image by encoding the local relationship between the intensity values of neighboring pixels. In the case of rice leaf disease diagnosis, LBP has been shown to effectively capture the changes in texture patterns caused by diseases. FD is another feature extraction technique that has been applied to rice leaf disease prediction. FD measures the complexity of an image by capturing the self-similarity of its shapes and patterns. In the context of rice leaf disease prediction, FD has been used to quantify the changes in the shapes of leaf images caused by diseases.

Lacunarity is a feature extraction technique that has been used in the analysis of complex images, including those related to rice leaf diseases. Lacunarity measures the degree of clustering and distribution of patterns in an image. In the case of rice leaf disease prediction, lacunarity has been used to quantify the changes in the spatial distribution of patterns caused by diseases. In terms of classifiers, several algorithms have been applied to the problem of rice leaf disease prediction, including support vector machines (SVMs), decision trees, k-nearest neighbors (KNNs), and artificial neural networks (ANNs). These classifiers have been trained on features extracted from rice leaf images using LBP, FD, and lacunarity. The results of these studies have shown that the combination of LBP and a classifier such as SVM or ANN can achieve high accuracy in the prediction of rice leaf diseases. In conclusion, LBP, FD, and lacunarity have been used for feature extraction in the diagnosis of rice leaf diseases. The results of various studies have shown that LBP and a classifier such as SVM or ANN can achieve high accuracy in the prediction of rice leaf diseases. Further research is needed to compare the performance of these feature extraction techniques and classifiers in the context of rice leaf disease prediction. To address the crop selection issues in practice of effective agriculture, Bhuvaneswari *et al.* [5] propose a model based on multi-label imbalanced classification problems using a meta-learning-based dynamic ensemble classifier selection approach. By shrinking the dataset using the VIKOR ranking method and leveraging meta-features to assess classifier competence, the classification complexity is reduced. The proposed work's novelty lies in determining classifier diversity within ensemble formations, ultimately improving the classification ability of the meta-learning-based dynamic ensemble model. The adoption of a customized voting strategy for the final ensemble output enhances its classification performance.

Previous studies have used digital image processing techniques such as Local Binary Patterns (LBP) and Color Histogram for feature extraction and various classifiers such as Random Forest and k-NN for disease classification in various crops such as corn and apple plants. In this study, we aim to build upon the previous research by combining LBP feature extraction and fractal features along with the ensemble classifier for the efficient and accurate detection and classification of paddy leaf diseases.

8.3   Methodology

In this research, we present a machine-learning-based framework for the detection and classification of paddy leaf diseases. The proposed framework consists of two main components, feature extraction and classification, as shown in Figure 8.2. In the feature extraction component, we use our proposed Improved Local Binary Pattern (ILBP) and Fractal features to extract the relevant features from the images of the paddy leaves. ILBP is used to capture the texture information of the paddy leaves while Fractal features are used to capture the geometrical properties of the images. In the classification component, we have used K-Nearest Neighbor, Support Vector Machine (SVM), and Adaboost Classifier to compare the performance of the proposed methodology, and the well-suited classifier will be used for classification purposes. The proposed framework is evaluated on a dataset consisting of images of paddy leaves affected by Rice Blast and Bacterial Blight. The results of the proposed framework are compared with other state-of-the-art methods and showed that our proposed framework outperforms other methods in terms of accuracy, precision, recall, and F1 score.

a) Pre-processing
The pre-processing step in image analysis involves various stages that help to bring the test image to the size, color, and quality of the images in the dataset. The steps include:

- Image resizing to match the dimensions of the training images
- Smoothing and converting the image to grayscale
- Noise filtering using a median filter

The purpose of pre-processing is to prepare the image for feature determination and extraction, making the image suitable for further analysis.

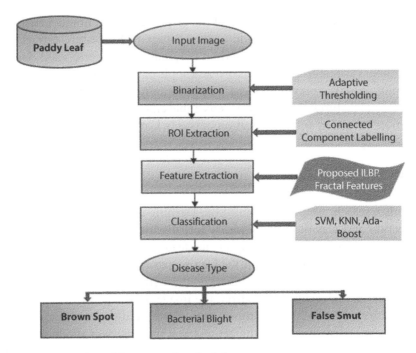

Figure 8.2 Overview of the proposed methodology.

b) Background Removal and ROI Segmentation
Paddy leaf image background removal and spot segmentation is a crucial step in the analysis of diseases in paddy plants. Adaptive thresholding and connected component labeling are two commonly used techniques for background removal in paddy leaf images. Adaptive thresholding adjusts the threshold value based on the local image intensity, while connected component labeling groups the connected pixels with similar intensity into distinct regions. These methods have proven to be effective in separating the paddy leaf from the background and enhancing the visibility of the disease symptoms, thereby improving the accuracy of disease identification.

c) Adaptive Thresholding
Adaptive thresholding is a common method used for image binarization, where an image is converted into a binary image, consisting of only black and white pixels. The main advantage of adaptive thresholding over global thresholding is that it can handle images with varying illumination conditions. The basic idea of adaptive thresholding is to divide the image into small sub-regions and calculate a threshold value for each sub-region

based on its local pixel intensity [6]. This threshold value is then used to binarize the pixels in that sub-region. To find t(x, y) for each pixel using the local adaptive threshold technique,

$$b(x, y) = \begin{cases} 0 & \text{if } I(x, y) \leq t(x, y) \\ 1 & \text{otherwise} \end{cases} \tag{8.1}$$

where b(x,y) is binary block, and the intensity of pixel is I(x,y).

d) Feature Extraction
For feature extraction, there are three techniques coined by the proposed framework: (i) Improved LBP, (ii) Fractal Dimension, and (iii) Lacunarity.

i. LBP Computation
Once the ROI is identified as explained above, the region is taken as mask and applied over the input image and then the grayscale ROI is extracted and given as input for TB-LBP computation. A powerful illumination invariant texture primitive is the local binary pattern (LBP) [7]. For texture description, the histogram of binary patterns generated across a region is considered. The operator characterizes each pixel in terms of its nearby pixels' relative gray levels. The value is set to one if the gray level of the surrounding pixel is higher or equal, otherwise to zero. After detecting LBP for each pixel, a histogram is constructed to represent the entire image as shown in Figure 8.3.

$$LBP_{r,n}(x, y) = \sum_{i=0}^{n-1} s(n_i - n_c)2^i, \tag{8.2}$$

$$s(x) = \begin{cases} 1, x \geq 0, \\ 0, otherwise. \end{cases} \tag{8.3}$$

Even though the basic LBP method acts as an efficient texture representation, it involves the extraction of 256 features, which leads to computational complexity. To overcome this issue, the LBP method proposed by Ojala [8] uses the idea of identifying uniformity in texture using transition. This method not only identifies uniformity, it also reduces the features from 256 to 59. In this method, the uniformity is observed when the obtained values are low and non-uniformity is explored when the value is high. In the

70	80	37
46	35	5
12	30	24

1	1	1
1	135	0
0	0	0

→ $(10000111)_2 = 135$

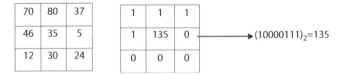

Figure 8.3 Illustration of LBP calculation.

proposed ILBP, an idea of reducing the features from 59 to 19 was attained by the following algorithm as shown in Figure 8.4.

In this algorithm, the maximum value produced in the rotational string will be converted and replaced by the center pixel. Hence, there must be only the probability of occurring 19 features as output. The proposed method is illustrated in Figure 8.5 with example. The transition pattern produced for the string 10111110 in Figure 8.3 is 11000011 (i.e., a bit wise shift operation is performed). Then, the rotational operation is performed for each bit.

In the illustrated diagram, the value 240 is replaced in the position of 135 in Figure 8.5. Hence, it is concluded that the proposed method can extract 19 unique features to represent the image. This indicates the dimensionality reduction and also the non-uniformity is represented by means of its highest value.

ii. Fractal Dimension and Lacunarity Estimation

Fractal dimension and lacunarity are two important parameters used for characterizing complex natural patterns such as those found in paddy leaf features. Fractal dimension is a measure of how much a pattern fills up space or how much space it occupies. It is a non-integer number that describes the degree of self-similarity of a fractal pattern at different scales. In other words, fractal dimension measures the degree of roughness or

Step 1	:	*Take the binary pattern computed from equation (2) as an input.*
Step 2	:	*Find the Transition binary pattern for the given input by one-bit shift operation.*
Step 3	:	*Rotate the Transition binary pattern in full clock-wise direction and compute the decimal value of each shifted pattern which produce 8 values as result.*
Step 4	:	*Find the maximum value obtained from the output produced at previous step and replace it to the center of the matrix.*

Figure 8.4 Proposed ILBP computation.

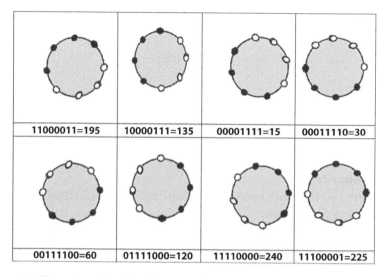

11000011=195	10000111=135	00001111=15	00011110=30
00111100=60	01111000=120	11110000=240	11100001=225

Figure 8.5 Illustration of rotation of transition bit.

complexity of a pattern. It can be computed using different methods, including the Box-Counting (BC) method and the Hausdorff dimension method. The formula for computing the fractal dimension using the box-counting method as shown in Equation 8.4 and Figure 8.6 illustrates the computation of FD using the log–log plot,

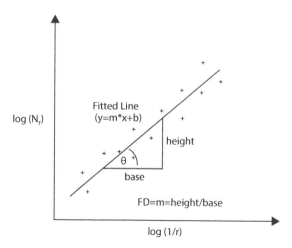

Figure 8.6 The log–log plot for the computation of FD.

$$FD = lim_{r \to 0} \frac{log(N_r)}{log\left(\dfrac{1}{r}\right)} \qquad (8.4)$$

where FD = fractal dimension,
   N_r = number of copies of A, scaled down by ratio "r"
   r = scaled ratio of A

iii. Lacunarity

Lacunarity, on the other hand, is a measure of how the distribution of gaps or lacunae within a pattern changes as the scale changes. It is a complementary parameter to fractal dimension and provides additional information about the spatial arrangement of features within the pattern [9]. Lacunarity is computed using the variance of the box-counting function, which measures the degree of variation in the number of boxes needed to cover the pattern at different scales. The formula for computing lacunarity using the box-counting method is shown in Equation 8.5.

$$L_r = \frac{\displaystyle\sum_M N_r^2 Q(N_r, s)}{\left[\displaystyle\sum_M N_r Q(N_r, s)\right]^2} \qquad (8.5)$$

where M is the size of the FD processed image
   $Q(N, s)$ is the probability of N in box size s,
   L_r is the lacunarity of box size s N_r and is computed using the BC method.

8.4   Results and Discussion

a) Dataset

The dataset used in this study contains 480 instances, of which 432 are in the training set and 48 are in the test set. These instances were sourced from the UCI repository and Kaggle, and were labeled with three categories. Figure 8.7 shows sample images from the dataset. To evaluate the performance of four classification algorithms, 10-fold cross-validation was performed on both the training set (90% of the dataset) and the test set (10% of the dataset). The accuracy results of the classification algorithms are presented in Table 8.1.

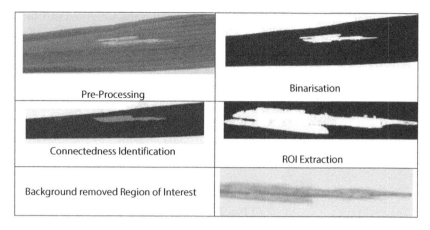

Figure 8.7 Sample output images.

Table 8.1 Sample features.

Image label	Fractal dimension	Lacunarity	ILBP
Normal leaf	1.10	0.21	170
	1.17	0.22	169
	1.15	0.19	143
Affected leaf	1.15	0.22	176
	1.20	0.15	178
	1.15	0.21	172

b) Feature Extraction

Sample features like Fractal Dimension, Lacunarity, and Improved LBP are shown in Table 8.1.

c) Classification

The proposed method tested with KNN at 10-fold cross-validation and k value is set to be 1. Then, the observed result is compared with another classifier in Table 8.3. Likewise, the Radial Basis Function (RBF) kernel is used to perform mapping of data from input space to feature space in SVM with 10-fold cross-validation. Finally, in the Adaboost algorithm, the number of iteration is set as 10 with seed value one and weighted threshold is set as 100. The Decision Stump has been chosen as base classifier and the predicted values for all the three databases are tabulated in Table 8.3 and graphical representation of the result is shown in Figure 8.4. Based on the result obtained, it is concluded that the proposed framework produces a good result on implementing the Adaboost algorithm over the three databases. The performance is measured with classification accuracy, which is defined as the total number of images correctly classified by the total number of images in the database and is defined in Equation 8.6.

$$Accuracy = \frac{(TP + TN)}{(TP + TN + FP + FN)} \times 100 \tag{8.6}$$

where TP is True Positive, TN is True Negative, FP is False Positive, and FN is False Negative.

Table 8.2 and Figure 8.8 present the evaluation of various performance measures such as TPR, FPR, Precision, Recall, F-Measure, and AUC for the four classification algorithms. Figure 8.8 depicts a comparison of their accuracy. It is evident that the ensemble approach consistently outperforms the other algorithms in detecting and classifying diseases, not just

Table 8.2 Performance comparison classifiers based on accuracy.

S. no.	Classifier	Parameter	Dataset	Accuracy
1	KNN	K = 1	Brown Spot	90.3%
		No. of Folds = 10	False smut	89.4%
			Bacterial Blight	87.5%
2	SVM	Kernel = RBF	Brown Spot	87.5%
		No. of Folds = 10	False smut	90.3%
			Bacterial Blight	89.5%
3	AdaBoost	Weight = 100	Brown Spot	94.7%
		Iteration = 10	False smut	95.7%
		Classifier = Decision Stump	Bacterial Blight	93.5%

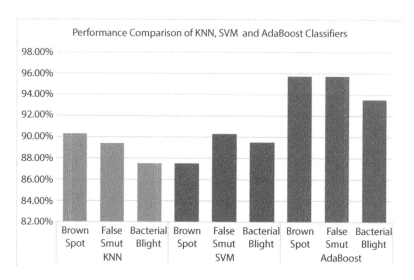

Figure 8.8 Performance comparison of KNN, SVM, and AdaBoost classifiers.

Table 8.3 Performance comparison with existing methods.

S. no.	Method	Dataset results	Results (accuracy in %)
1	Optimized DNN [9]	Private dataset of 650 images	92
2	Color+shape+texture feature +NN [10]	Private dataset of 30 images	92.5
3	Proposed method	Kaggle+UCI 480 images	94.6

in terms of accuracy but also in other performance metrics. We have compared our method with [10] and [11], to prove the robustness of our proposed method, and it is shown in Table 8.3.

8.5   Conclusion

The proposed framework for the detection and classification of paddy leaf diseases has shown promising results. The use of ILBP and Fractal features for feature extraction and the ensemble approach for classification have allowed for the development of a reliable and efficient system for the detection and classification of paddy leaf diseases. The results of the proposed framework have shown that it outperforms other state-of-the-art methods in terms of accuracy, precision, recall, and F1 score. The proposed framework can be used by farmers to detect the diseases affecting their paddy crops, and to take appropriate measures to control the diseases. The framework can also be used by researchers to develop new and more effective methods for the control of paddy leaf diseases. This research provides a step forward in the development of an efficient and effective system for the detection and classification of paddy leaf diseases and can contribute to the growth and sustainability of the paddy industry. In the future, to improve the precision of prediction result, we have planned to include deep neural network [12] and reinforcement-learning-based models [13] in our work.

References

1. *Paddy rice production worldwide*, Statista Inc., United States, 2022-04-01, https://www.statista.com/.
2. Barbedo, J.G., Arnal, L.V., Koenigkan, T.T.S., Identifying multiple plant diseases using digital image processing. *Biosyst. Eng.*, 147, 104–116, 2016.
3. Pantazi, X.E., Moshou, D., Tamouridou, A.A., Automated leaf disease detection in different crop species through image features analysis and one class classifiers. *Comput. Electron. Agric.*, 156, 96–104, 2019.
4. Cao, Y., Xu, H., Song, J. *et al.*, Applying spectral fractal dimension index to predict the SPAD value of rice leaves under bacterial blight disease stress. *Plant Methods*, 18, 67, 2022, https://doi.org/10.1186/s13007-022-00898-8.
5. Bhuvaneswari, Saravanan, P., Subramaniyaswamy, V., Meta learning-based dynamic ensemble model for crop selection. *Appl. Artif. Intell.*, 36, 1, 3627, 2022.
6. Wang, J., He, J., Han, Y., Ouyang, C., Li, D., An adaptive thresholding algorithm of field leaf image. *Comput. Electron. Agric.*, 96, 23–39, 2013.
7. Singh, V.P. and Srinivastava, R., Effective mammogram classification based on center symmetric-LBP features in wavelet domain using random forests', *Technol. Health Care*, 25, 4, 709–727, 2017.
8. Ojala, T., Pietikainen, M., Maenpaa, T., Multiresolution grayscale and rotation invariant texture classification with local binary patterns. *IEEE Trans. Pattern Anal. Mach. Intell.*, 24, 7, 971–987, 2002.
9. Plotnick, R.E., Lacunarity indices as a measure of landscape texture. *Landscape Ecol.*, 8, 3, 201–211, 1993.
10. Ramesh, S. and Vydeki, D., Recognition and classification of paddy leaf diseases using optimized deep neural network with Jaya algorithm. *Inf. Process. Agric.*, 7, 2, 249–260, 2020.

11. Mukherjee, M., Pal, T., Samanta, D., Damaged paddy leaf detection using image processing. *J. Global Res. Comput. Sci.*, 3, 10, 07–10, 2012.
12. Bhuvaneswari, Saravanan, P., Subramaniyaswamy, V., Feature fusion based deep neural collaborative filtering model for fertilizer prediction. *Expert Syst. Appl.*, 216, 119441, 2023.
13. Keerthana, S., Elakkiya, R., Belqasem, A., Srete, N., Subramaniyaswamy, V., Indragandhi, V., A systematic study on reinforcement learning based applications. *Energies*, 16, 3, 1512, 2023.

9

An Artificial Intelligent Methodology to Classify Knee Joint Disorder Using Machine Learning and Image Processing Techniques

M. Sharmila Begum[1], A. V. M. B. Aruna[1*], A. Balajee[2] and R. Murugan[3]

[1]Department of Computer Science and Engineering, Periyar Maniammai Institute of Science & Technology, Deemed to be University, Thanjavur, India
[2]School of Computer Science and Engineering, Faculty of Engineering and Technology, Jain Deemed to be University, Bangalore, India
[3]School of Computer Science and IT, Jain Deemed to be University, Jayanagar, Bangalore, India

Abstract

Disorder diagnosis at an earlier stage plays a vital role in the prediction and classification of samples in clinical support systems. Machine learning methods are gaining major importance in classifying the samples from the given data. In this chapter, a hybridization of isolation forest is proposed to classify the normal and abnormal knee joint signal samples. The number of features extracted from the raw data could lead to the complexity of the model. To reduce the feature burden of the model, this chapter proposes an artificial intelligence methodology to select the abstract set of features from the given raw data, and classification is done through Hybrid Isolation Forest (HIF). The chapter consists of three phases, starting from processing the raw images and extraction of the preprocessed dataset. The second phase identifies the abstract feature set using the statistical and regressive parameters. The third phase proposes a hybrid isolation forest method that integrates the probability distribution and isolation forest to perform the accurate classification of normal and abnormal data samples. The proposed AI model is tested on the knee joint disorder image dataset. The proposed model achieved the highest value of 98.31% in terms of accuracy, 97.13% in terms of sensitivity, and

[*]Corresponding author: aruna1418@pmu.edu

Elakkiya, R. and Subramaniyaswamy V. (eds.) Cognitive Analytics and Reinforcement Learning: Theories, Techniques and Applications, (167–188) © 2024 Scrivener Publishing LLC

95.33% in terms of specificity. The time complexity is also reduced for disorder classification by the proposed AI model.

Keywords: Knee joint disorder, diagnosis, hybrid isolation forest, feature selection, classification, and performance metrics

9.1   Introduction

Artificial intelligence and its subfields play a vital role in clinical disorder diagnosis at its earlier stages. Data analytics is used to analyze the statistical relationships that can be derived from the available data. The data are usually in the form of different types, which can be textual data, signal data, image data, and so on. Knee joint disorders are highly occurring disorders that are observed in the human skeletal system. There are a number of modalities that are adopted for diagnosing the various disorders occurring in the human knee joint. The modalities can be categorized into two major categories, namely, invasive and non-invasive types. The invasive modalities, including arthroscopy, open-knee surgery, etc., might lead to operational risks whereas non-invasive methods including Magnetic Resonance Imaging (MRI), and Computed Tomography (CT) are widely adopted for diagnosis. These non-invasive methods are good in their results but they are high in cost for continuous and periodic diagnosis. The major limitation observed with these mechanisms is that the samples are collected by the static placement of knee joints whereas the different levels of knee joint disorders usually respond highly during the dynamic movement of knee joints instead of their static placements.

One such recently emerged modality to diagnose the disorder levels during the static movement of knee joints is Vibroarthrography (VAG). The highest level of disorder in human knee joints is osteoarthritis, which occurs due to the continuous degeneration of knee joints caused by different levels of Chondromalacia Patella (CMP). The degeneration levels can be clinically graded as CMP 1, CMP2, and CMP3, and CMP4 leads to a higher level of disorder i.e., osteoarthritis. VAG is recorded as a signal that is collected from different individuals and the interpretation can be done through computational methods. The interpretation of such signals can be carried out in different ways that include various signal decomposition techniques to break down the signal into certain time stamps and the analysis can be performed. The alternate approach is to convert the VAG signals into a set of spectrogram images and the analysis can be performed by adopting statistical inferences from the images that identify the set of features that are useful for disorder diagnosis. The statistical interferences start with calculating various statistical features from the VAG signals including mean, variance, and standard deviation for the total population of data samples. Once the population parameters are calculated for the entire data samples, then the samples from each category were identified randomly, and their sample mean, sample standard deviation, and sample variance were calculated. By considering the sample proportion and population proportion, the hypotheses can be computed for individual samples considered for analysis. Then, a comparison between null hypotheses and alternative hypotheses was done with each of the samples to calculate the best set of features from the available data samples. Once the best set of features is identified, then the classification of normal and abnormal data samples is done through the proposed Hybrid Isolation Forest (HIF) that incorporates the Poisson distribution to the traditional isolation forest mechanism to classify the samples.

Once the classification is performed, then the obtained results of the proposed classification are compared with various existing algorithms including Linear Regression, K-Nearest Neighbor, Support Vector Machines (SVMs), Multilayer Perceptron (MLP), and Random Forest. The results of the proposed and the existing methods are compared with different performance metrics where the proposed HIF has the highest value of 98.31% in terms of accuracy, 97.13% in terms of sensitivity, and 95.33% in terms of specificity. The flow of the chapter continues in such a way that section II explains the various recent literature in the field of clinical data analysis and the impact of machine learning algorithms in disorder diagnosis. Section III defines the proposed methods used for disorder diagnosis that utilize various statistical parameters and their effectiveness in feature computation from the VAG signals. Section IV depicts the experimental results and their comparison with existing methods followed by the conclusion and references.

9.2   Literature Survey

Clinical data analysis using machine learning has become increasingly popular in recent years due to the vast amounts of data being generated in healthcare. In this literature survey, we will review some of the recent research papers that have focused on clinical data analysis using machine learning. Initially, Wu *et al.* (2010) proposed a computer-aided diagnosis (CAD) system for knee joint disorders using support vector machines (SVMs). The authors used a dataset of 200 knee-joint MRI images and achieved an accuracy of 95.5%. They also compared their model's performance with other machine learning algorithms and found that the SVM model outperformed them. Athavale and Krishnan (2020) describe an image-processing system for the detection of knee joint disorders. The system uses a combination of image segmentation, feature extraction, and classification techniques to identify abnormal regions in knee radiographs. Khalilia *et al.* (2015) describe a clinical predictive modeling workflow using open-source machine learning tools and provide examples of successful deployments.

Weins and Shennoy (2018) discuss the potential of machine learning in healthcare, the challenges faced in implementing machine learning models, and the ethical considerations that need to be taken into account. Solaries *et al.* (2020) provide a comparative review of deep neural architectures for clinical data analysis and highlight the strengths and weaknesses of each approach. Leukemia prediction in its earlier stages by incorporating the optimization strategies along with support vector machines classifier and better results were obtained by Santhakumar and Logeswari (2020). Binary classification for VAG signal samples using a hybrid feature selection mechanism was performed by Balajee and Venkatesan (2021). An initial attempt towards multiple classifications for knee joint clinical data are performed by Kreciz and Baczkowicz (2018) where the combination of various ML algorithms including Logistic Regression (for normal samples analysis), MLP, SVM, and Random Forest (for different abnormal samples analysis) are utilized for classifying the signal-based data samples.

Alphonse *et al.* (2021) proposed an advancement in a multiclass classification system to calculate the different levels of disorders in human knee joints where the different variants of the random forest algorithm provided better results when compared to the traditional ML algorithms. Thaventhiran *et al.* (2021) proposed the clinical analysis for lung cancer data where the target class is identified before processing the data and the samples are processed in such a way as to avoid the outliers using LSTM, which achieved the highest prediction accuracy of 93%. The effectiveness and importance of machine learning and classification systems in analyzing clinical data analysis were considered as a survey goal and a comparative study of various machine learning algorithms was done by Balajee and Venkatesan (2021).

A comprehensive review of various machine learning techniques used for knee joint disorder diagnosis is done by Kokkotis *et al.* (2020). The authors discuss the advantages and limitations of each method and compare their performance on different datasets. They also provide a detailed overview of the preprocessing techniques used to extract features from the knee joint images. The security mechanisms using different layers of networks were adopted and the features computed for classifying knee joint disorders were performed by Balajee *et al.* (2023). This mechanism improved the overall performance of VAG-based disorder analysis in human knee joints. The overall need for cognitive approaches to perform risk prediction at its earlier stages irrespective of the disease or disorders is highly demanding in the computational and clinical areas stated by Srivani *et al.* (2023). Li *et al.* (2023) applied various machine learning algorithms to perform predictive analysis using cognitive behavioral therapy where certain pre-predictors were set for the existing machine learning algorithms providing a better prediction rate when compared to the traditional ones.

Human assistance in formulating the data could lead to erroneous data collection, which can be avoided through the implementation of cognitive computing stated by Ahmed *et al.* (2017). There are numerous cognitive computing systems that are being developed to solve certain problems. IBM Watson for Oncology is a cognitive computing system providing better results in cancer treatment. In this system, a combinatory approach of natural language processing and machine learning is used to provide recommendations to clinicians for better treatment of lung and colorectal cancers (Somashekar *et al.*, 2017). A cognitive analysis of knee joint disorders using X-ray images and radiological assessment is done by Gornale *et al.* (2020). In this approach, the Kellgren-Lawrence grading concept is adopted for discriminating between two major levels of knee joint disorders including arthritis and osteoarthritis. The combination of image processing and information extraction from the knee joint disorder X-ray images using the cognitive system that consists of KL grading and HU-invariant movement calculation is done by Gornale *et al.* (2020), which provided better results for feature extraction and classification of knee joint disorders.

9.3   Proposed Methodology

The initial phase of the work starts with obtaining a spectrogram image from the available signal data. The problem we had considered for analysis (i.e., VAG Signal Analysis) consists of non-linear signals that have the tendency to change in terms of different time stamps. The architecture of the proposed method is shown in Figure 9.1.

a) Signal Preprocessing and Feature Computation

In order to evaluate the frequency's time evolution, spectrogram image computation is very much essential. Initially, we discriminated the VAG signals into a uniform set of segments by fixing the time length as 5 seconds. Short Fourier Transform is obtained through windowing each of the segments and the spectrum computation from the overall segmented VAG signal samples. The decibel representation of individual spectrum power had to be displayed through the spectrogram representation of raw signal samples. Once the representation of the VAG signal through spectrogram is available, the next stage is to derive the frequency domain, time domain, and a set of statistical features from the VAG signal through various signal processing and machine learning packages available in Python. An example of time–frequency representation obtained through processing the normal and abnormal VAG signals through the Short Fourier Transform (SFT) mechanism is shown in Figure 9.2, which shows the absolute variation of signal samples between the normal and abnormal samples.

The single time-domain frequency cannot handle the classification of all VAG signals; hence, we extracted the signals in five different domains including the statistical, time-domain, Hjorth's, non-linear, and signal complexity features. A total of 25 features are available in the preprocessed VAG dataset. The initial set of features computed through the available raw signal data is the statistical features set. A total of 11 statistical features

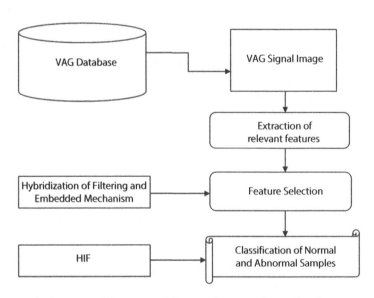

Figure 9.1 Architecture of the proposed feature selection and HIF classification.

were available in the initial set, which starts with identifying the correlation between the signals. The correlation between the signals that are broken down with different time stamps is calculated using Equation 9.1.

$$(f * g)(\tau) \triangleq \int_{-\infty}^{\infty} f(t - \tau)g(t)dt \tag{9.1}$$

Here, f and g represent two signals with continuous time stamps. Correlation represents the convolution of one VAG sample with functional inversion of the other.

$$\sum_{n=-\infty}^{\infty} |x(n)|^2 \tag{9.2}$$

Equation 9.2 represents the energy of each and every VAG signal. Energy is used to denote the area that comes under the squared magnitude for each and every VAG sample that is being processed through the signal transformation mechanism. The change in signal from its input stage to the output that changes for every frequency dimension can be calculated using homogeneity as mentioned in Equation 9.3.

$$\text{Homogeneity} = \sum_N \frac{(observed - expected)^2}{expected} \tag{9.3}$$

The consecutive statistical features can be calculated from the VAG signal samples, which are denoted in the following VAG signal samples. Equation 9.4 denotes the mean calculated from the VAG signal.

$$\mu = \frac{1}{N} \sum_{i=0}^{N-1} x_i \tag{9.4}$$

The standard deviation of the signal is calculated from the number of signals considered for analysis as mentioned in Equation 9.5.

$$\sigma = \sqrt{\frac{\sum_{i=1}^{N} (x_i - \mu)^2}{N}} \tag{9.5}$$

One of the most important features that measure the power spectral density is entropy, which can be calculated as mentioned in Equation 9.6.

$$\text{Entropy} = -\sum_{i=1}^{N} p(x_i) log_2 p(x_i) \qquad (9.6)$$

Root mean square is used to calculate the mean average of a sample over a certain period of time as mentioned in Equation 9.7

$$\text{RMS} = \sqrt{\frac{1}{N} \sum_{i=1}^{N} x_i^2} \qquad (9.7)$$

We calculated the deviation experienced in every signal from its original mean value through variance, the squared value of the standard deviation.

$$\text{Variance} = \sigma^2 \qquad (9.8)$$

Smoothness is considered an essential factor in categorizing the normal and abnormal VAG signals that can be calculated through the probabilistic function represented through Equation 9.9.

$$\text{Smoothness} = g(x) = x^2 sinsin\left(\frac{1}{x}\right) for\ x > 0 \qquad (9.9)$$

The maximum displacement achieved by the signal in the spectral analysis is obtained by calculating the kurtosis of the signal.

$$\text{Kurtosis} = n * \frac{\sum_{i}^{n} (y_i - \underline{Y})^4}{\sum_{i}^{n} (y_i - \underline{Y}^2)^2} \qquad (9.10)$$

The difference in VAG signal distribution is calculated through the skewness value obtained for the signal, which is obtained through Equation 9.11

$$\text{Skewness} = \sum_{i=1}^{N} \frac{(x_i - \mu)^3}{(N-1) * \sigma^3} \tag{9.11}$$

Irregularity factor (IF) for normal and abnormal VAG signals is computed. The normal signals have differentiation in IF but the abnormal IF is almost similar for all pathological samples. The traversal waveform length is also computed for the signals with distinct values for signifying the abnormality level indications of the signals. RSZON is calculated based on the number of samples and the time displacement kept constant at 0.5. The formulation adopted to calculate RSZON is given below:

$$m_0 = \sqrt{\sum (n^2)}^{\left(\frac{1}{t}\right)} \tag{9.12}$$

RSSON and RSFON are obtained by calculating the second- and fourth-order derivatives for the signals that are considered for analysis.

$$m_2 = \frac{\sqrt{\sum (d1)^2}^{\frac{1}{t}}}{n-1} \tag{9.13}$$

$$m_4 = \frac{\sqrt{\sum (d2)^2}^{\frac{1}{t}}}{n-1} \tag{9.14}$$

Sparseness is the time-domain feature representing the relevant and non-redundant nature of a VAG signal. It can be calculated using Equation 9.15.

$$\text{Sparseness} = \frac{\sqrt{\left|(m_0 - m_2) * (m_0 - m_4)\right|}}{m_0} \tag{9.15}$$

In the above equation, m_0, m_2, and m_4 denote the root squared zeroth-order, second-order, and fourth-order derivatives. The irregularity factor (IRF) measures the smoothness between the different derivatives of a VAG signal.

$$IRF = \frac{m_2}{\sqrt{m_0 * m_4}} \tag{9.16}$$

The Wavelength Ratio (WLR) feature is computed to calculate the cumulative length of the waveform over a different time using the following formulation.

$$WLR = \frac{\sum |d_1|}{\sum |d_2|} \tag{9.17}$$

Here, d_1 and d_2 denote the higher-order derivatives calculated from the VAG signal samples considered for analysis.

One of the essential features for non-linear data is Hjorth's features, which segregates signal performance into three subcategories: activity, mobility, and complexity. Activity features computed for abnormal VAG signals have high variability since different levels of abnormality are observed in VAG signals.

$$Activity = \sigma^2(x) \tag{9.18}$$

We can compute mobility along with log value to track the drifted velocity of the signals.

$$Mobility = \sqrt{\frac{\sigma^2(x^|)}{\sigma^2(x)}} \tag{9.19}$$

Complexity for signals is denoted as a set of irregular or non-linear data conveyed on separate time instances from different time series.

$$\text{Complexity} = \frac{Mobility(x^{|})}{Mobility(x)} \qquad (9.20)$$

Non-linear features, along with earlier computed features, and a set including entropy for measuring the spread of power density for the spectrum had been calculated.

$$ShEn = -\sum_{l=l_{min}}^{N} p(l)lnlnp(l) \qquad (9.21)$$

We had calculated permutation entropy to identify the various combinations performed with the signal data as one of the non-linear features.

$$PeEN = -\sum_{1}^{k} p_k log(p_k) \qquad (9.22)$$

We had computed three complexity features for the VAG signal, namely, approximate, range, and sample entropy. In addition, the approximation for the spectral power density spread over the entire data had been identified using AP-EN.

$$ApEn(m,r,N) = \left[\varnothing^{m}(r) - \varnothing^{m+1}(r)\right] \qquad (9.23)$$

Here, m denotes the length of the data segmentation that has been considered for analysis. r denotes the similarity of the data, and N denotes the number of samples. The maximum and minimum range in complex signals are calculated using Range entropy as follows:

$$RaEn(\alpha) = \frac{1}{1-\alpha}\left(\sum_{f} e_f^{\alpha}\right), \alpha > 0, \alpha \neq 1 \qquad (9.24)$$

The common spread among the VAG data is given through Sample entropy calculation.

$$SampEn(m,r,N) = ln\left[\varnothing^{m}(r) - \varnothing^{m+1}(r)\right] \qquad (9.25)$$

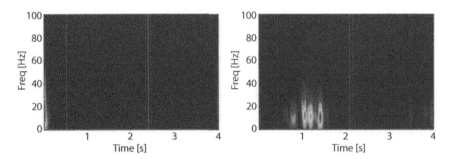

Figure 9.2 SFT time–frequency spectrogram representation of normal and abnormal VAG samples.

b) Feature Selection

Once the features are computed and preprocessed from the raw VAG signals, it is essential to process the dataset towards the feature selection mechanism in order to calculate the relevance of features towards the target class. The feature selection mechanism has a number of approaches to select the relevant subset features that could lead to better classification. The three major approaches to feature selection are the filter approach, wrapper approach, and embedded approach.

Filter in feature selection is termed as an interrelationship within the features from a feature set. In VAG analysis, let us consider the age and year of diagnosis as an interrelated feature to cross-check the nature of a sample that mainly focuses on the relevance of features. The filter approaches in ML and statistics include Information Gain (IG), which represents the maximum information obtained from a single feature, Chi-Square test for evaluating the usefulness of a feature, correlation coefficient to determine how the useful features are interrelated, and so on.

The useful information from the features is the driving force of wrapper approaches in feature selection. According to our work on VAG signal analysis, a wrapper can be considered as the most important and useful feature that can be identified from the feature vector and can be quoted as a wrapper feature that can classify the given samples. ML provides strategies such as Recursive Feature Elimination (RFE), sequential algorithms used for feature selection, and GA approaches for feature selection.

Embedded approaches have a certain common relevance with Wrapper, but the building blocks of embedded approaches are obtained while constructing the feature model, whereas in a wrapper, that information is identified earlier. In our earlier work on binary classification of VAG signals, we utilized the embedded approach as a feature selection criterion (Balajee and Venkatesan, 2021). Regularization methods and decision trees can be utilized as an embedded approach for selecting the best feature set.

The alternate approach apart from selecting the features is to extract the relevant set of features using certain statistical and machine learning mechanisms.

c) Feature Extraction Approaches

Instead of ranking or ordering the set of available features and reducing the features, an alternate and efficient approach followed by multiple classifications and predictions is feature extraction. In this mechanism of feature engineering, the features are mapped to higher-dimensional representations, and the combinatory set of features is obtained by the algorithms. There are many feature extraction algorithms available where PCA and LDA have a high impact on clinical data analysis.

PCA is a feature extraction and dimensionality reduction method in which statistical methods are induced for extracting the features belonging to higher-dimensional space. In PCA, the standardization of the given data is performed first from which the covariance matrix had to be formed for identifying the interrelationship in data. Eigenvalue and eigenvectors are composed for identifying the principal components of the data. Now, the extracted feature has a number of principal components from which the classification can be performed.

LDA is an efficient feature extraction and classification technique while working with samples that have binary classes. Here, the features are incremented in each iteration for identifying the class of the sample or data. In our VAG analysis, the computed features with respect to different domains are combined together and the randomness of the data is identified using the LDA method. The outcome of the LDA will be almost similar to the class of the labeled data sample. As mentioned earlier, in our work, LDA is utilized as a supervised feature extraction strategy for mapping the given data with higher-dimensional space.

d) Classification

In machine learning and data analytics, the essential mechanism for identifying the underlying pattern or class of the sample is to classify the samples based on the interpretation provided through the sample. There are a number of classification algorithms that are being developed for handling various problems based on the nature of the problems. The classification algorithms can be categorized into three major areas based on the data availability, that is, supervised, unsupervised, and reinforcement classification algorithms. The samples with labeled data (where the class discriminations are known) can be handled by supervised classification algorithms whereas samples with unlabeled data (where the class discriminations are unknown) can be handled with unsupervised classification algorithms. The reinforcement mechanism acts as a trial-and-error mechanism to identify the class discriminations of the samples. Since the classes of the available data are known in the dataset considered for our problem, we process the data samples through a supervised machine learning algorithm.

The earlier research works carried out by Balajee *et al.* (2023) claimed that the Random Forest and its different variations had provided better results in classifying the normal and abnormal data samples. The overall performance results provided by the RF algorithm can still be improved by isolating the erroneous parameters that increase the complexity of the ML model. Thus, we implemented the Hybrid Isolation Forest algorithm that combines the feature selection mechanism along with the isolation forest so that the overall complexity is reduced and the performance of the classification system is also increased. The mechanism of the proposed Hybrid Isolation Forest (HIF) along with its statistical parameters considered for the implementation is discussed in the following section.

e) Hybrid Isolation Forest

Specific data that deviate from normal are usually denoted as anomalies in analytics. In clinical data analytics, detecting the anomaly is the most important aspect to discriminate the classes, which could improve the overall performance of the proposed model. The deviated data from the normal representations are also termed outliers, which could be highly related to the anomaly data. In real-world scenario, the global anomalies highly differ from the normal data points, which means the distribution of certain data points would be far from the statistical parameters derived for the entire dataset. The traditional isolation forest algorithm works similar to that of the random forest by ensembling the number of decision trees together to formulate a decision whereas the algorithm differs in dealing with the samples in that it considers every data point as an unlabeled one and it also has an assumption that the criterion to claim the given sample is an anomalous one will be very unique and fewer in the dataset. In our proposed Hybrid Isolation Forest (HIF), we induced the concept of hypothesis analysis to the traditional isolation forest algorithm, which improvised the overall performance of the proposed method. The initial phase of the proposed algorithm is to identify the null hypothesis and an alternative hypothesis for the considered data sample. Then, hypothesis testing was performed to determine whether the statement (i.e., the type of hypothesis) about the value of a population parameter should or should not be rejected. In our approach, we adopted the concept of normal distribution, and the assumption of the population variance is considered as our hypothesis. Once the null and alternate hypothesis is framed, then the test statistic about the hypothesis is performed using the following equation:

$$t_{n-1} = \frac{x - \mu_0}{\dfrac{s}{\sqrt{n}}} \qquad (9.26)$$

Here, the term n denotes the sample size, $\underline{X}$ denotes the sample mean, μ_0 denotes the null hypothesis, and S denotes the standard deviation of the overall sample data. The considered statistic has the t-statistic with $n-1$ degrees of freedom. Since we had considered two hypotheses, i.e., for determining the difference in terms of normal and abnormal samples, the test statistic can be extended for two mean values as follows:

$$t = \frac{(X_1 - X_2) - (\mu_1 - \mu_2)}{\sqrt{\dfrac{S_p^2}{n_1} + \dfrac{S_p^2}{n_2}}} \qquad (9.27)$$

where

$$S_p^2 = \frac{(n_1 - 1)s_1^2 + (n_2 - 1)s_2^2}{((n_1 + n_2) - 2)} \qquad (9.28)$$

Here, n_1 and n_2 denote the degrees of freedom and s_1 and s_2 denote the standard deviation derived by the parameters available in each of the hypothesis. By considering the above criteria, the proposed Hybrid Isolation Forest is developed and it was compared with other classification algorithms used for binary classification in the following section.

9.4   Experimental Results

In the proposed work, we utilized the VAG signal data obtained from the research group headed by Professor Rangayyan, University of Calgary, Canada.

The dataset contains 89 samples that have 38 normal and 51 abnormal data samples in it. The signal data with normal and abnormal signals are shown in Figure 9.3. The dataset was processed with signal-processing methods as mentioned in section 9.3.

The total number of features available in the dataset is 25, which belongs to five different categories of features. The hybrid two-stage feature selection method uses a statistical approach as a filtering method, i.e., mutual information is computed between the features, and based on the dependency between the components, a relevant feature subset is obtained. In this approach, we had given our entire feature vector, which has 89 samples and 25 features, as an input to the filter method. The outcome of the filter method had given a feature subset that has 15 attributes, including energy, entropy, smoothness, kurtosis, root mean square, RS0 norm, RS4 norm, irregularity factor, sparseness, mobility, complexity, permutation entropy, approximate entropy, range entropy, and sample entropy. Stage 2 uses an embedded approach to identify the crisp set of features from the relevant feature subset. L1 regularized logistic regression is utilized as an embedded

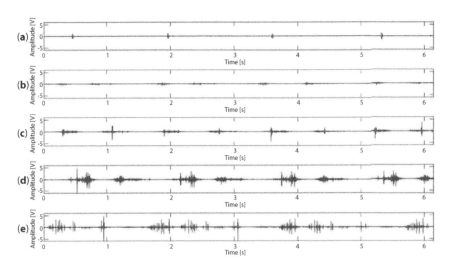

Figure 9.3 Comparison between normal (a) and abnormal (b, c, d, e) VAG signals.

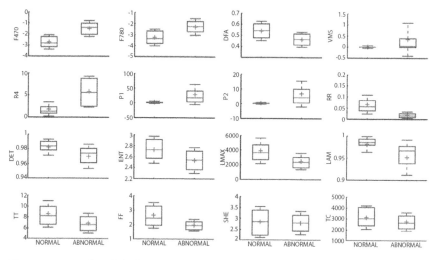

Figure 9.4 Box plot of the features selected.

method to which the 89 samples along with 16 specified features are given as an input.

By applying the proposed feature selection mechanism, we reduced the number of features to 16, which reduced the complexity of the data for performing the classification. The impact of features in discriminating the normal and abnormal samples is shown in Figure 9.4. The embedded process had given nine crisp features as an outcome of the regression analysis, which include entropy, kurtosis, root mean square, RS0 norm, RS4 norm, sparseness, mobility, permutation entropy, and range entropy. The impact of feature computation over the classification phase is analyzed through the graphical representation, which denotes the spread of the data over the feature space. This varies between the normal and abnormal data samples considered for analysis. The graph representation of data over the computed feature space for normal and abnormal models is shown in Figures 9.5 and 9.6. The classification result obtained by the proposed Hybrid Isolation Forest (HIF) algorithm along with the existing methods is shown in Table 9.1 where the proposed algorithm outperformed the existing methods in terms of accuracy, specificity, and sensitivity. The graph representation of data for normal models in the sample set of computed features that consists of homogeneity, mean, standard deviation, irregularity factor, wavelength ratio, and mobility is considered as a cluster of features that belong to different domains and are utilized for representation.

From Table 9.1, it is seen that the proposed HIF algorithm achieved the highest accuracy of 98.31%, highest sensitivity of 97.13%, and highest

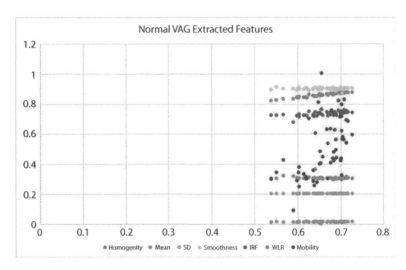

Figure 9.5 Data spread in feature space for standard samples.

specificity of 95.33%, which outperformed the existing methods. Thus the overall performance of the AI-induced knee joint disorder identification obtained better results when compared to all the other diagnosis mechanisms.

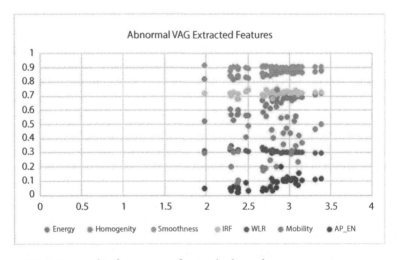

Figure 9.6 Data spread in feature space for standard samples.

Table 9.1 Comparison of performance metrics for the proposed HIF with existing methods.

S. no.	Classifier	Accuracy (%)	Sensitivity (%)	Specificity (%)	Time complexity (seconds)
1	Linear regression	70.83	72.67	69.78	1.489
2	KNN	71.13	72.46	70.33	1.210
3	Multilayer perceptron	82.38	85.55	80.29	0.935
4	SVM	86.66	87.31	84.32	1.185
5	HIF (Proposed)	98.31	97.13	95.33	0.889

9.5   Conclusion

The major focus of this chapter is to develop an artificial intelligence method to diagnose knee joint disorder using vibroarthrographic signals. The initial phase constructed a spectrogram image from the raw signal that was useful in identifying the appropriate set of features. Once the features are computed using the statistical parameter normal distribution, the 16 appropriate features are identified as the best features. The final phase developed a Hybrid Isolation Forest (HIF) classifier to categorize the samples into normal and abnormal categories with the highest performance results. The proposed feature selection and classification methods were compared with existing methodologies including Linear Regression, KNN, Multilayer Perceptron, and SVM, where the proposed HIF outperformed in terms of accuracy, specificity, sensitivity, and time complexity. The combination of statistical parameters along with the AI methods is useful in creating a novel machine learning model that is useful to handle a real-world problem. In the future, the signals can be broken down with fixed time stamps using deep learning algorithms so that the prediction of clinical and accidental disorders can be done at much earlier stages.

References

1. Athavale, Y. and Krishnan, S., A telehealth system framework for assessing knee-joint conditions using vibroarthrographic signals. *Biomed. Signal Process. Control*, 55, 101580, 2020 Jan 1.
2. Khalilia, M., Choi, M., Henderson, A., Iyengar, S., Braunstein, M., Sun, J., Clinical predictive modeling development and deployment through FHIR web services. *AMIA Annu. Symp Proc. 2015*, 2015, 717, 2015.
3. Wiens, J. and Shenoy, E.S., Machine learning for healthcare: On the verge of a major shift in healthcare epidemiology. *Clin. Infect. Dis.*, 66, 1, 149–53, 2018 Jan 1.
4. Solares, J.R., Raimondi, F.E., Zhu, Y., Rahimian, F., Canoy, D., Tran, J., Gomes, A.C., Payberah, A.H., Zottoli, M., Nazarzadeh, M., Conrad, N., Deep learning for electronic health records: A comparative review of multiple deep neural architectures. *J. Biomed. Inform.*, 101, 103337, 2020 Jan 1.
5. Santhakumar, D. and Logeswari, S., Hybrid ant lion mutated ant colony optimizer technique for Leukemia prediction using microarray gene data. *J. Ambient Intell. Hum. Comput.*, 12, 2965–73, 2021 Feb.
6. Balajee, A. and Venkatesan, R., Machine learning based identification and classification of disorders in human knee joint–computational approach. *Soft Comput.*, 25, 20, 13001–13, 2021 Oct.
7. Kręcisz, K. and Bączkowicz, D., Analysis and multiclass classification of pathological knee joints using vibroarthrographic signals. *Comput. Methods Programs Biomed.*, 154, 37–44, 2018 Feb 1.
8. Alphonse, B., Rajagopal, V., Sengan, S., Kittusamy, K., Kandasamy, A., Periyasamy, R., Modeling and multi-class classification of vibroarthographic signals via time domain curvilinear divergence random forest. *J. Ambient Intell. Hum. Comput.*, 23, 1–3, 2021 Feb.
9. Thaventhiran, C. and Sekar, K.R., Target projection feature matching based deep ANN with LSTM for lung cancer prediction. *Intell. Autom. Soft Comput.*, 31, 1, 495–506, 2022 Jan 1.
10. Balajee, A. and Venkatesan, R., A survey on classification methodologies utilized for classifying the knee joint disorder levels using vibroarthrographic signals. *Mater. Today: Proc.*, 80, 3240–3243, 2021 Jul 26.
11. Kokkotis, C., Moustakidis, S., Papageorgiou, E., Giakas, G., Tsaopoulos, D.E., Machine learning in knee osteoarthritis: A review. *Osteoarthr. Cartil. Open*, 2, 3, 100069, 2020 Sep 1.
12. Balajee, A., Murugan, R., Venkatesh, K., Security-enhanced machine learning model for diagnosis of knee joint disorders using vibroarthrographic signals. *Soft Comput.*, 21, 1–1, 2023 Feb.
13. Srivani, M., Murugappan, A., Mala, T., Cognitive computing technological trends and future research directions in healthcare–A systematic literature review. *Artif. Intell. Med.*, 23, 102513, 2023 Feb.

14. Li, F., Jörg, F., Merkx, M.J., Feenstra, T., Early symptom change contributes to the outcome prediction of cognitive behavioral therapy for depression patients: A machine learning approach. *J. Affect. Disord.*, 334, 352–7, 2023 Aug 1.

15. Ahmed, M.N., Toor, A.S., O'Neil, K., Friedland, D., Cognitive computing and the future of health care cognitive computing and the future of healthcare: The cognitive power of IBM watson has the potential to transform global personalized medicine. *IEEE Pulse*, 8, 3, 4–9, 2017 May 16.

16. Somashekhar, S.P., Sepúlveda, M.J., Norden, A.D., Rauthan, A., Arun, K., Patil, P., Ethadka, R.Y., Kumar, R.C., Early experience with IBM watson for oncology (WFO) cognitive computing system for lung and colorectal cancer treatment. *J. Clin. Oncol.*, 35, 15, 8527–8528, 2017.

17. Gornale, S.S., Patravali, P.U., Hiremath, P.S., Cognitive informatics, computer modeling and cognitive science assessment of knee osteoarthritis in radiographic images: a machine learning approach, in: *Cognitive Informatics, Computer Modelling, and Cognitive Science*, pp. 93–121, Academic Press, Amsterdam, The Netherlands, 2020 Jan 1.

18. Gornale, S.S., Patravali, P.U., Hiremath, P.S., Automatic detection and classification of knee osteoarthritis using hu's invariant moments. *Front. Rob. AI*, 7, 591827, 2020 Nov 16.

19. Wu, Y., Krishnan, S., Rangayyan, R.M., Computer-aided diagnosis of knee-joint disorders via vibroarthrographic signal analysis: A review. *Crit. Rev. Biomed. Eng.*, 38, 2, 201–224, 2010.

Part III

ADVANCEMENTS IN COGNITIVE COMPUTING: PRACTICAL IMPLEMENTATIONS

10

Fuzzy-Based Efficient Resource Allocation and Scheduling in a Computational Distributed Environment

Suguna M.[1]*, Logesh R.[2] and Om Kumar C. U.[1]

[1]*School of Computer Science and Engineering, Vellore Institute of Technology, Chennai, India*
[2]*Centre for Advanced Data Science, Vellore Institute of Technology, Chennai, India*

Abstract

The sharing of resources results in better performance at a cheaper cost, making grid and cloud computing an attractive choice for high-performance computing. It enables the cooperative use of a variety of resource allocation distributed virtual machines. The objective is to reduce the cost while achieving maximum resource utilization, minimal reaction time, and a load that is evenly distributed among all the resources used in a grid and cloud. Fault-tolerant scheduling is an essential step to maintain backup overloading that is utilized for independent initiatives to cut costs. Integration, security framework interoperability, and participant trust relationships are the primary issues. It provides the modified fuzzy-logic-based self-adaptive resource allocation and scheduling method. Security requirements are transformed into a fuzzy collection using empirical membership functions. With this strategy, scheduling success rates are higher and performance against resource failures is robust. In cloud computing, a few techniques are already in use for resource allocation. The novel resource allocation technique based on hybrid fuzzy clustering will be proposed after an analysis of these current cloud resource allocation schemes. The suggested method ingeniously reserves resources whose power substantially exceeds the needs of present jobs for use in the future while allocating adequate resources to tasks that precisely satisfy their resource needs.

*Corresponding author: suguna.m@vit.ac.in

Elakkiya, R. and Subramaniyaswamy V. (eds.) Cognitive Analytics and Reinforcement Learning: Theories, Techniques and Applications, (191–202) © 2024 Scrivener Publishing LLC

Keywords: Resource allocation, scheduling, distributed computing, fault tolerant, fuzzy logic

10.1   Introduction

Due to the heterogeneous nature of the resources and their varying security regulations, security systems in a grid context are exceedingly complex and difficult to implement. The computer resources are distributed across a variety of security domains and platforms. Data integrity, confidentiality, and information pricing are the focus areas for security requirements [1, 7]. There are currently several security frameworks and standards accessible. People and organizations favor the option that best fits their surroundings. Security frameworks are not interchangeable or reproducible. A crucial function of cloud computing is load balancing. To prevent processing delays and resource overcommitments, it must be connected into both grid and cloud systems. There are three types of load balancing: centralized, decentralized, and hierarchical. Grid middleware has challenges in providing its users with a manageable distributed, secure, stable, and high quality-of-service system due to grids' decentralized approach, flexibility and heterogeneous architecture, and goal of being a general-purpose system [2, 3, 9]. Information communication is necessary when distributing workloads among several resources. In distributed computing systems, many different scheduling can be used in a cloud setting. Fuzzy-based algorithm avoids allocating strong resources to straightforward and medium-scale projects or weak resources to difficult large-scale activities, both of which may result in resource waste and task scheduling failure [13]. The suggested algorithm is anticipated to operate more effectively than existing similar algorithms and to be more robust. The key feature of a new many-to-many stable matching algorithm is that it effectively matches servers with virtual machines that have different resource requirements. This work will offer a fresh method for cloud computing. Understanding how grid and cloud-based hosting can be utilized to overcome some of the current limitations and increase the capabilities of numerous types of apps is not difficult [10, 12].

Resource sharing is one of the main aspects of grid computing. However, this resource sharing will lead to a few security difficulties because each service layer implementation needs security interoperability [4]. Building trust among those taking part in the grid ecosystem is really challenging. In the face of uncertainty, fuzzy set theory aims to describe one that can mathematically define a fuzzy set [6, 16]. Fuzzy allows for membership degrees ranging from 0 to 1. A membership function converts the universe's constituent parts into values. It is relatively simple to make a judgment if the system's output is fuzzy as opposed to a crisp value. Defuzzification is the process of turning output into a crisp value [5]. The load balancing methodology employs the neighbor-based and cluster-based load balancing techniques. It combines load balancing and fault tolerance scheduling. Both the backup and primary replication strategies are passive [7, 8].

10.2   Proposed System

For resources not allocating to simple and medium-scale projects or weak resources to challenging large-scale activities, the recommended method may prevent resource waste and task scheduling failure in grid. The proposed approach is projected to function more reliably and efficiently than the currently used, comparable techniques. The main characteristic of a novel many-to-many fuzzy stable matching method is that it successfully matches servers with virtual machines that require various amounts of resources. This work will present a novel approach to grid and cloud computing, as shown in Figure 10.1.

The fuzzy parameter trust level of the grid environment is calculated using the trust level of the resource and speed. The membership functions and inference rules will be used by the fuzzy interference process to map the values. The defuzzification technique is used to calculate the replication number for each process. A group of computing nodes are taken into consideration in this experiment. Each computer node is assumed to have one or more processors. Each scheduler has a network connection to the processor's node. The user node is assigned to the processor and delivered

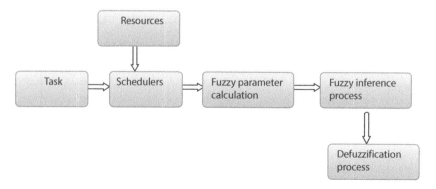

Figure 10.1 System architecture.

to the scheduler for distribution to the processors. With this approach, fuzzy logic is used.

Figure 10.2 depicts [18, 19] resource allocation using cloud virtual machine. Backup overloading is the simultaneous scheduling of backups for several primary systems. It effectively makes use of the processing time. For fault-tolerant scheduling, the primary and backup approaches are used. The backup is activated when the primary fails. Backups for all nodes in a system do not necessarily have to be planned after primaries for all nodes. Failures in projects and processors are discovered using the fail-signal and acceptance test fault detection mechanism. The Load Balancing Model will choose a neighboring resource for offloading projects from among the p number of surrounding resources that each resource (ci) maintains (LNSeti).

$$\alpha = \frac{TD_{ik}}{TD_{inearest}} \tag{10.1}$$

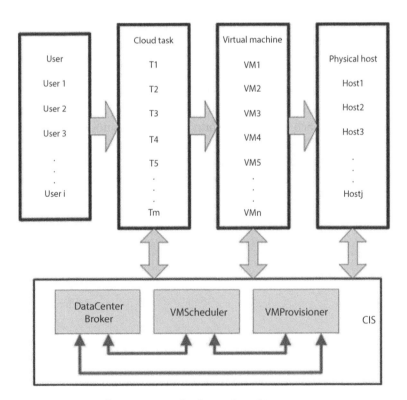

Figure 10.2 Resource allocation using cloud virtual machine.

where Transfer Delay [TD_{ik}] represents the delay in moving the resource ci to the target ck. The transfer delay from resource ci's closest resource to itself is indicated by the symbol $TD_{inearest}$. There will be an independent score for each resource while multiple projects are submitting to the same resource.

$$h(c_q) = \sum_{i=1}^{k(q)} s(i) \qquad (10.2)$$

where the number of projects being submitted to the resource cq is represented by k(q). The efficiency of each resource can be modified as O (k).

$$\text{Effectiveness} = O\ (k).\ \text{Efficiency} + h(cq) \qquad (10.3)$$

Fuzzy parameter is calculated using the formula below [4]:

$$\overline{SD} = \frac{\sum_{i=1}^{n}\left(SD_i \times \sum_{j=1}^{q_i}\frac{e_{ij}}{q_i}\right)}{\sum_{i=1}^{n}\sum_{j=1}^{q_i}\frac{e_{ij}}{q_i}} \qquad (10.4)$$

Security Demand [SD] stands for each node's security need, while e stands for anticipated execution time. The hosts that meet the security assurance condition are represented by q. The equation is used to calculate the grid environment's trust level.

$$\overline{TL} = \frac{\sum_{j=1}^{m}TL_j \times p_j}{\sum_{j=1}^{m}p_j} \qquad (10.5)$$

TL stands for the trustworthiness of each resource, and P stands for speed. Security error ratio is an additional factor. It utilizes the equation to calculate.

Table 10.1 Fuzzy inference rule.

S. no.	SD	$\overline{SD}$	SEi (throughput)	Ki (TENDENCY)
1	High	Medium	Medium	Medium
2	High	Medium	Medium	High
3	Low	Medium	Medium	Medium
4	Low	Medium	Low	Low
5	High	Medium	Low	Low
6	High	High	Low	Medium

$$SE_i = \frac{\overline{TL} - SD_i}{SD_i} \qquad (10.6)$$

Fuzzy inference processes of the parameters in this module are mapped into the range [0,1] using membership functions. The membership functions are used to obtain five levels of SD, SE, and K, namely, very low, low, medium, high, and very high. Thus, the membership degrees for the SD, SE, and TL bars will be created. The grid environment performs better thanks to the fuzzy inference rules. The parameters throughput and tendency [14] are used, along with fuzzy parameter rules, to create fuzzy inference rules shown in Table 10.1.

Each node derives its replication number throughout the defuzzification phase. Defuzzification is the process of reducing a set of fuzzy values to a single crisp value. The process is the opposite of fuzzification.

10.3   Experimental Results

After values for the number of resources are assigned in the NetBeans environment, the scheduling of jobs begins. Each resource's and job's security demand and trust level develop dynamically. A membership function is used to generate membership degrees for the SD, TL, throughput, and failure tendency. Apply fuzzy operations to map the input space to the output space using the fuzzy rule set. Six fuzzy inference rules are currently being generated. For fault-tolerant scheduling, the number of replicas is calculated using the defuzzification method. The system is provided with a set of resources as input. The user can enter the number of grid users, regional GIS users, resource users, and job users here. The resources are distributed at random based on how many GIS are present. One of the routers receives the resources on a dynamic basis. For the allocation, an LNset and LPset concept is employed [17].

a) Calculating Allocations and Loads for GIS

The method for configuring the environment variables on the machines is shown in Figure 10.5. The number of grid users, regional GIS users, resources, and occupations can all be added here by the user. The user will be taken to a new window with the scheduling information after clicking the submit button after entering the parameters.

Figure 10.3 (a and b) displays the calculation of load in each resource. Here, according to the load, the parameter throughput and failure tendency is calculated. Grid and cloud computing is an evolution of distributed computing. It deals with resource trade that is efficient and based on customer requirements. Additionally, the data center is home to a considerable number of diverse resources [14, 15]. The virtualization capabilities of cloud computing make it different from other previously supplied kinds of computing by hiding the heterogeneity of the resources using cloudsim. The Cloudlet class now has two additional variables, which are expected completion time and expected bandwidth. The values of these variables can be accessed and changed in four different ways. Fuzzy logic is implemented using MATLAB. The distribution of resources is quantitatively modeled in this work utilizing fuzzy logic techniques. This figure demonstrates how the scheduler for an infrastructure as a service cloud model uses fuzzy logic. The Cloud Lease Scheduler is widely used in situations where dynamic load circumstances are accessible. Although the resources are spread out, there are no techniques. Based on the resources that are

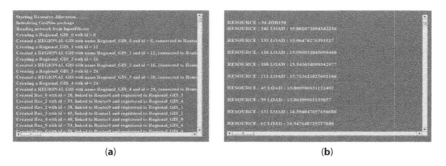

(a)   (b)

Figure 10.3 (a) Allocating to GIS, (b) Calculating load.

available, such as the many CPU cycles on another node and the available primary memory on one node, the administrator can make decisions. Logical resources could be made available to provide virtual resources and other services. One of the most crucial criteria for evaluating the effectiveness of fault-tolerant systems is throughput [11, 13]. Throughput is:

$$\text{Throughput} = n/\ Tn \tag{10.7}$$

An execution matrix is created by taking the resources and project ID listed in Table 10.2 into account. Here, a value is dynamically created for each project to show how long it will take to complete the work using a specific resource. The scheduling is done in accordance with the amount of time created for each project and the demand on each resource. Time Average response time per request measures how long it takes to send a response after receiving a request for resource allocation. The recommended response time is calculated as the total of the request-dependent times. The suggested way can state that the make span for each project is the ART.

Table 10.3 and Figure 10.4 present the PDMinRc, which only considers a single replica with a main and backup strategy. No security settings are considered, but the graph still displays better results. The X-axis in this graph reflects task load, which ranges from 0 to 90, and the Y-axis represents the average reaction time, which ranges from 200 to 1000 ms, with a 4.3% success rate in scheduling.

Table 10.2 Expected execution time matrix.

Project ID/ Resources	R1	R2	R3	R4	R5	SD level
1	17.086	3.28	15.716	30.874	40.845	Low
2	8.443	30.083	19.889	39.973	48.89	High
3	14.179	36.028	45.95	23.216	2.376	Low
4	47.068	30.509	12.954	5.714	12.905	High
5	4.316	3.282	44.529	46.936	45.584	High
6	4.23	33.43	22.275	48.015	21.579	High
7	5.078	8.79	26.172	14.458	39.959	High

Table 10.3 Average response time comparison.

Number of resources	PDMinRc	MFSARS
10	684	305
20	608	502
30	834	524
40	586	597
50	894	392
60	682	376
70	924	595
80	911	482
90	936	491

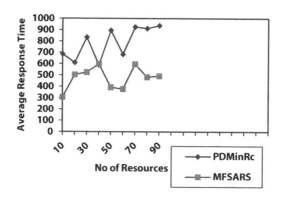

Figure 10.4 Average response time comparison.

Table 10.4 and Figure 10.5 depict the comparative performance show task load, which ranges from 0 to 90, while the Y-axis shows resource scheduling success rate, which ranges from 50 to 80. Based on the findings, the proposed method has a greater rate of success than PDMinRc.

Figure 10.6 (a and b) presents this statistic, which is represented by a replication cost graph and shows how Fuzzy Logic-Based Secure can reduce the amount of processor time required for backup. According to the results, our suggested algorithm performs better than the current approach. Reliability cost is the proportion of jobs that can be regained

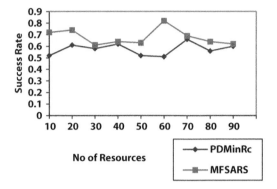

Figure 10.5 Scheduling success rate comparison.

Table 10.4 Scheduling success rate comparison.

Number of resources	PDMinRc	MFSARS
10	0.52	0.72
20	0.61	0.74
30	0.58	0.61
40	0.62	0.64
50	0.52	0.63
60	0.51	0.82
70	0.66	0.69
80	0.56	0.64
90	0.60	0.62

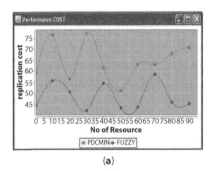

(a)

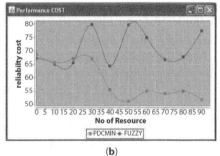

(b)

Figure 10.6 (a) Replication cost graph, (b) Reliability cost graph.

following a failure. These data demonstrate the error tolerance capacity of fuzzy logic-based secure.

10.4   Conclusion

A fuzzy logic-based technique takes project security requirements and resource trust levels into account when making fuzzy inferences. The trust relationship between the project and the resource is guaranteed by the security assurance condition. In fault-tolerant scheduling, a predetermined number of replicated copies of the projects use an excessive number of hosts or resources. The number of replicated projects is calculated using the suggested algorithm. In this study, resource distribution is mathematically modeled using fuzzy logic methods. This diagram illustrates how fuzzy logic is implemented in the scheduler for an infrastructure as a service cloud paradigm. The suggested approach offers improved scheduling success rates and strong performance against resource failures. The distribution of resources is quantitatively modeled in this work utilizing fuzzy logic techniques. The Cloud Lease Scheduler is widely used in situations where dynamic load circumstances are accessible. Although the resources are scattered, there are no available methods for making provisions to fulfill the customer's outstanding request. Based on the resources that are available, such as the many CPU cycles on another node and the available primary memory on one node, the administrator can make decisions.

References

1. Foster, I. and Kesselman, C. (Eds.), *The Grid: Blueprint for a Future Computing Infrastructure*, second ed, Morgan Kaufmann Publishers, 2004.
2. Cody, E., *et al.* Security in grid computing: A review and synthesis. *Decs. Support Syst.,* 44, 4, 749–764, 2008.
3. Suri, P. K. and Singh, M., An efficient decentralized load balancing algorithm for grid. *2010 IEEE 2nd International Advance Computing Conference (IACC),* IEEE, 2010.
4. Rajasekaran, S. and Vijayalakshmi Pai, G.A., *Neural networks, fuzzy logic and genetic algorithm: Synthesis and applications (with cd).* PHI Learning Pvt. Ltd., 2003.
5. Amoon, M., A fault tolerant scheduling system based on check pointing for computational grids. *Int. J. Adv. Sci. Technol.,* 48, 115–24, 2012.
6. Malarvizhi, N. and Uthariaraj, R.V., Hierarchical status information exchange scheduling and load balancing for computational exchange scheduling and

load balancing for computational grid environments. *Int. J. Comput. Sci. Netw. Secur.*, 10, 2, 177–18, 2011.

7. Suguna, M. and Sharmila, D., Efficient virtualization technique for cloud to achieve optimal resource allocation. *J. Inf. Sci. Eng.*, 33, 859–869, 2017.

8. Suguna, M. and Sharmila, D., Resource allocation and scheduling using an optimized time driven approach in grid environment'. *J. Comput. Theor. Nanosci.*, 13, 11, 8627–8632, 2016.

9. Zheng, Q., Tham, C.-K., Veeravalli, B., Dynamic load balancing and pricing in grid computing with communication delay. *J. Grid Comput.*, 6, 239–253, 2008.

10. Zheng, Q., Veeravalli, B., Tham, C.-K., On the design of fault-tolerant scheduling strategies using primary-backup approach for computational grids with low replication costs. *IEEE Trans. Comput.*, 58, 3, 380–393, 2008.

11. Suguna, M. and Sharmila, D., Project task scheduling in grid computing using optimization algorithm. *Int. J. Appl. Eng. Res. (IJAER)*, 10, 6, 14029–14040, 2015.

12. Han, C.-C., Shin, K.G., Wu, J., A fault-tolerant scheduling algorithm for real-time periodic tasks with possible software faults. *IEEE Trans. Comput.*, 52, 3, 362–372, 2003.

13. Suguna, M. and Sharmila, D., Heuristic task workflow scheduling in cloud using spot and on-demand instances. *J. Comput. Theor. Nanosci.*, 15, 8, 2640–2644, 2018.

14. Om Kumar C.U.and Bhama, P.R.K.S., Fuzzy based energy efficient workload management system for flash crowd. *Comput. Commun.*, 147, 225–234, 2019.

15. Dhakal, S., Hayat, M.M., Pezoa, J.E., Yang, C., Bader, D.A., Dynamic load balancing in distributed systems in the presence of delays: A regeneration theory approach. *IEEE Trans. Parallel Distrib. Syst.*, 18, 4, 485–497, 2007 April.

16. Deepa N, K. and Nithya, L.M., Fuzzy logic based job scheduling in computational grid with minimum communication and replication cost. *ICETS Proceedings on International Conference on Engineering Technology and Science*, pp. 446–451, 2014.

17. Om Kumar, C., Tejaswi, K., Bhargavi, P., A distributed cloud-prevents attacks and preserves user privacy, in: *2013 15th International Conference on Advanced Computing Technologies*, 2013.

18. Suguna, M., Fuzzy logic controller based IoT for smart irrigation system, in: *2021 International Conference on Advancements in Electrical, Electronics, Communication, Computing and Automation (ICAECA)*, IEEE, pp. 1–5, 2021 October.

19. Anupama, K.C., Shivakumar, B.R., Nagaraja, R., Resource utilization prediction in cloud computing using hybrid model. *Int. J. Adv. Comput. Sci. Appl.*, 12, 4, 2021.

11

A Lightweight CNN Architecture for Prediction of Plant Diseases

Sasikumar A.[1], Logesh Ravi[2], Malathi Devarajan[3], Selvalakshmi A.[4] and Subramaniyaswamy V.[5*]

[1]*Department of Data Science and Business Systems, Faculty of Engineering and Technology, SRM Institute of Science and Technology, Kattankulathur, Tamil Nadu, India*
[2]*Centre for Advanced Data Science, Vellore Institute of Technology, Chennai, India*
[3]*School of Computer Science and Engineering, Vellore Institute of Technology, Chennai, India*
[4]*Vellore Institute of Technology, Chennai, India*
[5]*School of Computing, SASTRA Deemed University, Thanjavur, India*

Abstract

Detecting maize crop diseases accurately is a difficult task that farmers face throughout the maize development and production stages. TensorFlow will be used for Deep Learning and the Convolutional Neural Network (CNN) architecture. Existing CNN-based solutions have been demonstrated to be less accurate and incompatible with most datasets and applications. We developed a real-time, automated system for detecting plant illnesses based on the CNN architecture. Using a reinforcement learning model on synthetic and real-world images, the proposed system is trained to categorize the input into sickness. The proposed architecture provides high accuracy, a lightweight model, and speedy processing. The experimental result shows that the proposed architecture accurately predicted types of plant diseases, and the system performance did not decline over time. It can be used for real-time disease detection on plants, and botanists and academics may quickly access it for research.

Keywords: Convolutional neural network, deep learning, plant diseases, lightweight CNN, reinforcement learning

Corresponding author: vsubramaniyaswamy@gmail.com

Elakkiya, R. and Subramaniyaswamy V. (eds.) Cognitive Analytics and Reinforcement Learning: Theories, Techniques and Applications, (203–222) © 2024 Scrivener Publishing LLC

11.1   Introduction

Recent years of dramatic population development have raised the need for agricultural products, resulting in a significant expansion of farming. Crop yield productivity needs to double in output by 2050 to keep up with the growing population's need for food, bio-energy, and meat products. Key crop yields need to increase by 2.4% per year to meet this objective, but they are currently only rising by about 1.3% annually [1]. However, if this requirement is met, the ecosystem will suffer from losing biodiversity and increasing greenhouse gas emissions. Optimizing the utilization of resources like water and soil to enable high-yield crops is essential because traditional agricultural production is economically and environmentally viable.

Early warning technologies are urgently needed in precision farming because producing horticulture products requires a sufficient effective use of assets on an economic, ecological, and social level. Tomato farming faces major difficulties due to the variety of ailments, pests, and other abnormalities that might affect these crops globally [2]. For example, seeding structures, tracking, robotics, gathering, choice, and treatment are some of the technological domains where attempts are made to present solutions with different methods from which the manufacturing setting's outcomes and circumstances are continuously enhanced. Recent years have seen an acceleration in the growth of free equipment, which has sparked the creation and use of low-cost agricultural surveillance equipment with the ability to use AI and processing of images. For instance, Osroosh *et al.* present the development and execution of a system for tracking using RGB and heat images, executed on the Raspberry Pi 3, which controls function in actual and challenging settings [3]. This pattern continues to rise, but it offers accessibility, dependability, and movement as some of its main difficulties. Raspberry Pi is utilized to construct an inexpensive surveillance network for agricultural uses to achieve greater adoption. Smart systems provide a surveillance ecosystem with several sensors and connectors to build connection capabilities in inexpensive gadgets for agricultural surroundings.

In agricultural settings, the identification of diseases is primarily accomplished by skilled inspection of sight. However, it is limited by human limits, such as exhaustion brought on by lengthy workdays or the chance of making a mistake when carrying out a task that can become extremely repetitious. Due to the intricacy and numerous visual and contextual aspects accompanying these tasks, creating tools for their automation is a subject of ongoing attention.

Particularly in issues related to classification, the adoption of deep learning models has increased dramatically [4, 5]. It has also been employed in agriculture for projections, scenario rebuilding, picture improvement, identification, and restoration. At the same time, it has lately been utilized for crop categorization, identification of anomalies, water anxiety, categorization and choice of goods, and disease or pest identification in a particular sector of gardening [6]. It is characterized as a collection of structures in images from multiple sources, including infrared light heat, visible spectrum, multispectral, hyperspectral, and green fluorescence detectors. Making choices for developing a model for illness detection becomes more difficult due to this range of options. The amount of bandwidth and quantity of bands that will be used for consideration during developing a data matrix are two distinguishing features that distinguish these sensors.

Most detection techniques used in current studies are created in multispectral images that span the optical bandwidth. In contrast, most reinforcement learning use RGB images due to the volume of data available to previous studies for reproduction and enhancement of suggestions, while the data from multispectral detectors offering additional details, the lack of open datasets, and the price of machinery are a restriction to their use.

In related studies, most warning algorithms are created utilizing RGB or multispectral images, with most of the effort concentrated on deep learning in the RGB domain. This results from the volume of data already gathered through earlier studies. While multispectral sensor information offers more details, their greater usage is constrained. This is important in developing nations due to the absence of publicly available datasets and the expensive hardware cost. Due to its higher reliability and accessibility of equipment, neural networks have thus recently been preferred to machine learning techniques for classifying illnesses in datasets with agricultural origins.

Studies on deep learning and plant illnesses also demonstrate transfer learning using cutting-edge networks AlexNet and VGG16, producing findings with greater accuracy than those attained by artificial intelligence classifiers. Additionally, Brahimi et al. propose saliency maps for recognizing locations on leaf surfaces with specific patterns and categorization systems for 10 tomato illnesses using reinforcement learning from established models like AlexNet and GoogleNet [7]. Concerning device designs, other initiatives have been made to use more up-to-date designs, such as ResNet and Xception, with superior outcomes over earlier designs. Using the NVIDIA Jetson TX1 technology to carry out training created a categorization model for 10 categories of tomato photos based on the AlexNet framework. Lastly, data for five agricultural products and 17 lessons were gathered to images in real circumstances, providing three categorization algorithms centered on the ResNet50 design and creating a tool for handheld devices that allows the use of the studies. The combined dataset has been improved to develop additional portable uses of the developed models.

Plant disease identification is a prerequisite for efficiently and precisely preventing plant disease in a complex environment. The rapid expansion of smart farming has resulted in the digitization and data-driven detection of plant disease, allowing for better decision support, analysis, and planning [8, 9]. Deep learning algorithms are still being developed, but they have substantially contributed to identifying plant diseases, giving a reliable instrument with extremely exact results. The number and quality of labeled data used for training significantly impact the accuracy of deep learning models. The input is a plant ailment that has to be diagnosed, and the output is the application's suggestion of which disease the maize plant has been diagnosed with the help of machine learning algorithms. The purpose of developing this application using machine learning algorithms is to considerably benefit people in identifying and diagnosing plant diseases before flowering. To this end, we developed a transfer learning model to improve disease plant detection and offer a viable alternative to manual prediction and diagnosis in laboratories.

The rest of this chapter consists of the following sections. The challenges of precision agriculture are described in Section II. Section III describes a lightweight CNN of the existing architectures. Section IV presents the proposed CNN model for plant diseases using the transfer learning model. Section V illustrates applications of lightweight CNN models for sample images and their results. Lastly, Section VI summarizes the overview of the chapter.

11.2   Precision Agriculture

The practice of collecting, analyzing, and reacting to numerous intra- and inter-field variation inputs for contemporary agriculture is known as precision agriculture [10]. According to common terms, precision agriculture is a method of agricultural operations that uses innovation to track, measure, and analyze the demands of particular fields and crops. In practice, precision agriculture and conventional farming are very different. In conventional agriculture, farmers evenly distribute pesticides, fertilizers, and irrigation throughout all of their fields at predetermined intervals and rates, following the standard guidelines for the area. Even within a single area, there are always variations in natural, physical, and chemical features. Inputs are overused in fertile areas and underused in less fertile ones when fields are treated uniformly without considering their natural variations. This wasteful use of land, water, gasoline, fertilizers, and insecticides increases the cost and ecological effect.

By using Variation Frequency Adjustment, precision agriculture maximizes input while focusing on the inherent variances in the field. Geographic information systems (GIS) and plant lifespan utilizing GPS and satellite imagery are required to collect comprehensive geographic information spanning farms and locations necessary to implement VRA [11, 12]. Precision irrigation networks, production tracking, and mapping instruments and data managing platforms are some of the precise agricultural tools used. Precision agriculture uses sophisticated predictive, forecasting, and regulatory statistics algorithms to analyze the data to apply economic, environmentally friendly, and durable contemporary agricultural remedies. Figure 11.1 illustrates the challenges of precision agriculture.

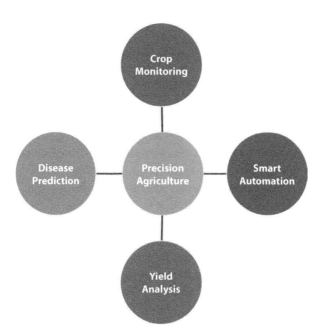

Figure 11.1 Challenges of precision agriculture.

a) Disease Prediction

Creative methods that provide higher yield and durability for indoor and outdoor farming are needed to increase agricultural output. Farmers need precise technology to collect and understand data to manage crop growth more effectively, reduce losses due to unfavorable weather conditions or contagious pests, and facilitate the return of expenditures. Plant diseases are thought to be responsible for $220 billion in annual global crop losses. To estimate the disease intensity and crop loss, it is essential to analyze the evaluation of disease and pathogens. Numerous agroecosystems suffer due to the widespread use of pesticides like bactericides, fungicides, and nematodes to manage plant diseases. A precise and trustworthy approach is required in plant disease evaluation to boost plant disease diagnosis and intensity calculation.

Different sensing devices and methods have been created for the early identification of disease signs when they are evident to the unaided eye, thanks to advancements in information and technology [13]. Using the idea of temporal variation, targeted protection of plants provides an alternative way of controlling plant diseases. For this reason, there currently exist three primary sensing systems: close-range systems, like ground-based or portable sensors; middle-range systems, like cameras attached to self-sufficient drones (UAVs); and far-range systems, like piloted aircraft or satellite-based detectors. Various tailored methods and methodologies, such as deep learning and artificial intelligence, have been developed for processing sensor data.

b) Crop Monitoring

Over the projected period (2021–2026), crop monitoring equipment is anticipated to see a CAGR of 12.6% in the precision agriculture industry [14]. Tracking crops has grown more essential for maximizing the potential of geographic information systems and location-specific crop control techniques as technology has evolved since its introduction. The market is expanding due to government efforts, the merging of IT and farming, and falling sensor costs. Over the projection period, it is projected that an increasing number of telemetry uses in agriculture would fuel the need for monitoring crops. Tracking gadgets that use the Global Navigation Satellite Systems to demonstrate the location of the machinery for administration reasons are part of telemetry solutions. These solutions are observed to be adopted most frequently in regions with high labor costs and low land costs. Soon, new growth prospects are anticipated due to the rising market for UAVs and the current efficacy of precision agriculture in raising the production of important crops.

Additionally, economic growth is being boosted by the actions made by different authorities to meet the rising food consumption. For example, the US Ministry of Food and Agriculture, the space agency, and NOAA support agricultural precision farming by creating cutting-edge GPS navigational systems. Precision Farming Development Centers (PDFCs) also receive a 100% cost subsidy from the Indian government. However, a lack of awareness and hefty startup expenses could limit the market's expansion. Additionally, due to poor entry, industrialized countries now employ technology the most.

c) Smart Automation

Almost all farming tasks are optimized through automated software and devices, often known as smart farming. Farmers can use gadgets, computerized systems, and extensive data collection to boost agricultural yields and revenue using less water, fertilization, and pesticides. Automated technology is being utilized to simplify manual operations previously completed by humans, allowing farmers to use their energy and time better. A few instances of automated technology that is frequently used in fields include [15, 16]:

Robotics: Robots can perform routine tasks like pruning, gathering, and sowing. Machines can work more accurately than people because they can make the most of technology like reinforcement learning and machine vision to increase productivity.

Devices for Smart Agriculture: Autonomous agricultural tractors can complete automated tasks using detectors, accurate GPS, and artificial intelligence.

Sensors: Sensors keep an eye on the soil's climate, transpiration, and wetness. Farmers can precisely forecast the best dates for growing and fertilizing using the data gathered. To enhance the number of crops established, monitors can also be utilized to map out the most effective sowing path.

With the capacity to communicate data or instructions back to automated machinery, farmers can change settings using automated control systems without being present on the farm. A prime illustration is a system of governance for watering. Farmers can establish a trigger for switching watering on or off utilizing robust sensors. This may be used in other machinery as well. An alert can be delivered immediately to the farmers by text or mail if a fuel condition sensor indicates that the oil in an apparatus has to be replaced. Previously, watering was set up using a straightforward timer that turned on mechanically at predetermined intervals.

Yet, monitoring and control of timed watering is frequently not possible. Thanks to robust gadgets, farmers can turn on, turn off, and change settings for controlled drip irrigation using a mobile device. Farmers can access the data in real time and make decisions without having to be on-site, thanks to sensors placed in key watering system locations.

The main generating source for most electrical lines, often a hydropower or fossil energy power station, only provides energy in one direction. The energy production scheme used with these systems requires that the energy generated by these source materials surpass the requirement by a significant amount to ensure an uninterrupted power supply because electricity production is controlled via responses from the transformer, but there is no knowledge input from the customer side. Determining flaws in these systems and fixing them is likewise a time-consuming task. In addition, as renewable energy technologies become more affordable, consumers now produce their energy and receive supplies from the major utility. The term "smart grids" refers to using ICT technologies to increase the observability of existing and newly installed grids, enable decentralized electricity production at both the consumer and utility ends, and add self-healing abilities to the grid. At various grid locations throughout the supply lines until the consumer, real-time power data are provided to utilities. Smart grids enable the integration of various energy sources and the network's self-healing [17, 18] to ensure an uninterrupted supply and better control of power production using predictive models generated from gathered consumer data.

d) Yield Analysis

Predicting crop yields is one of the most difficult problems in agriculture [19]. It is crucial to make international, regional, and local choices [20]. Agricultural, soil, climatic, ecological, and other characteristics predict agricultural yield. For crop forecasting, assistance with choice models is often used to extract important crop attributes. Surveillance (perceiving technology), managerial systems, fluctuating rate methods, and reactions to inter- and intravariability in agricultural systems are the main focus areas for precision farming. Precision farming can improve agricultural yield and quality while having fewer adverse effects on the surroundings.

Understanding the accumulated impacts on water and fertilizer deficits, pests, illnesses, the effect of yield variation, and various other field variables across the growing period is made easier with the assistance of crop yield simulators. The ability to forecast produce using "simpler" statistical techniques or choices that promote structures that are currently in use consequently, as well as potential applications of machine learning, is made possible by farm and *in situ* reports combined with current databases. The latter has the benefit of having the ability to handle numerous variables in perpetuity in both space and time; i.e., big data records developed with accuracy leadership instruments and data gathering features can be used in the fields of weather, the internet, and soil-related data, such as characterizing different kinds of plants.

11.3   Related Work

The existing systems are based on CNN and are found to be less accurate and almost incompatible with most datasets and applications. Despite promising prediction results, these traditional approaches must be more accurate and efficient. The existing systems are simple and effective but are extremely vulnerable to impact. Moreover, state-of-the-art methods can predict outputs about certain parts of the plant only, while some pest infestations may go completely undetected.

Although they are still in their infancy, ML-based solutions for farming are already promising. For example, it is possible to classify diseases from images using well-liked CNN implementations for various plants with various diseases [20]; to predict future insect damage, it is possible to recover interactions among weather information and pest presence using Long Short Term Memory channels and to detect insects on leaves using features extraction and deep learning tools. Farmers now have access to commercialized smart farming equipment and services that heavily rely on reinforcement learning. Here are a few instances. Plantix is an app agricultural assistant developed by PETA [21]. It aids in detecting plant diseases using machine learning and deep learning algorithms.

A startup firm in Switzerland named Gamaya provides a wide range of smart agricultural services based on the analysis of photos taken by drones linked to IoT devices. Another business that provides IoT-based soil testing and satellite imaging-based crop monitoring products is The Asian iFarmer. The huge tow-behind glyphosate sprayer See & Spray, created by California-based Blue River Technology, employs machine and deep learning-based algorithms to instantaneously identify and locate weed growth (in real time), adapting glyphosate to the exact locations discovered instead of to the entire field.

A few related studies can be discovered in the literature, but most concentrate on conventional ML techniques. In [22], a correlation of ML algorithms for estimating the yield of corn and soy crops is introduced. Scientific studies from the last 10 years for forecasting the beginning of disease at an initial or presymptomatic stage are evaluated and classified. The potential of using distinct ML techniques in agricultural production is addressed, but most of the present work is focused on conventional ML algorithms.

The proposed system is an automated, real-time system for plant disease detection that uses Densenet 121 architecture [23]. It has been trained on synthetic and real images using reinforcement learning to classify the input of diseases. The advantages of the proposed system are security, interpretability, high accuracy, lightweight model, and fast processing, which are desired in plant research, and its performance does not deteriorate with time. It could be used for real-time detection of plant diseases and can easily be used by botanists and researchers for research. Another application of the system can be in the field of pest management.

In this stage, the project's viability is assessed, and commercial plans are presented with a broad strategy. The feasibility study also includes some cost estimations. Some knowledge of the major criteria for the system is required in order to ensure that it is practical. The analysis' three main factors to be taken into account are as follows:

1. The system's price should fall within the company's budgetary constraints.
2. The system's implementation should take less time than the company's given period.
3. The system's quality must meet the criteria established by the business.

This research is being done to create an economical system. The corporation has limited resources that it may use to construct the system [24]. As a result, the designed system had to be reasonably priced. The majority of the technologies were already in use. However, it was necessary to buy specialist software and equipment. The system ought to be light and practical. This implies that there should not be any intensive server-side processing.

Additionally, the database must have the fewest number of records possible. This is due to the requirement for transferring the client-side copies of the server-stored data. Therefore, limiting the amount of data delivered will lessen network traffic. By monitoring whether or not they are willing to try out the new system, users can gauge the level of acceptability of the system. They are requested to share their opinions and recommendations regarding any issues they encounter during the trial period [25]. Their willingness to do so suggests that they find the system acceptable.

a) A Lightweight CNN Model for Precision Agriculture

The possibility for immediate execution on mobile devices exists for light-weight CNN models. The models above function well in identifying the presence of plant diseases because CNNs have strong feature extraction capabilities. By altering the design's framework, several lightweight works draw inspiration from the ILSVRC design [26]. Dilation of convolution, an origin component, and an altered AlexNet framework were all introduced by Zhu *et al.* [27]. The global pooling layers in place of an entirely interconnected plane and dilatation in the convolution layer have been implemented to preserve the computational burden. On 600 pictures of cucumber disease in six categories, an accuracy of 94.6% has been recorded. As an alternative to deep CNNs, Li *et al.* [28] concentrated on shallow networks. To extract picture features, they introduced two shallow networks containing the first four layers of the VGG-16 architecture. For the categorization job, Linear SVM and random forest approaches were utilized [29, 30]. The grain, apple, and grape-related illnesses are accurately predicted by the kernel SVM with 0.94 accuracy. On the PlantVillage group, Agarwal *et al.* [31] built a straightforward 6-layer CNN model for tomato disease detection and discovered an accuracy of 98.4%.

Experts have recently used the focus process to enhance CNN model accuracy. Ramamurthy *et al.* [32] added an awareness method to identify plant diseases at the top of the residue phase. They achieved 98% efficiency on four categories of tomato plants using the PlantVillage dataset. By using data enhancement, they were able to raise the sample size from 5,392 to 95,999. Enhancing the data includes trimming, enlargement, and varying the contrast level.

ShuffleNet designs featuring squeeze-and-excitation (SE) units were utilized by Tang *et al.* [33] to incorporate concentration mechanisms into their model for the diagnosis of grapevine disease. Using the revised ShuffleNet v1 approach, researchers achieved 99.14% efficiency. Brainstorm modules and SE blocks were also used to provide excellent accuracy results on tomato species from the PlantVillage dataset and the aubergine dataset [34]. A deep CNN framework with a self-focus block was utilized by Agila [35]. The AESCD9214 and MK-D2 datasets obtained accuracy rates of 95.33% and 98.0%, respectively. Depthwise separable convolution with thick units and a system for attention was employed by Chen *et al.* [36]. The model is effective with maize species. The MobileNet v2 model was employed with the concentration method to detect rice illness. On 10 classes from the PlantVillage and Rice datasets, they obtained accuracy rates of 99.67% and 98.48%, respectively. Most compact models discussed here have been evaluated solely for effectiveness on a tiny portion of the PlantVillage dataset or a limited number of crop species. PlantVillage employed the most numerous plant species, with five species and 10 groups.

In short, early techniques for identifying plant diseases relied on image processing and conventional artificial intelligence techniques like SVM, Multiple Layered Perceptron, Naive Bayes, and KNN. A recent study has focused on deep CNNs, and researchers have offered various potential solutions by employing reinforcement learning techniques and conventional CNN structures. In identifying plant diseases, lightweight CNN models with limited layers have also been investigated. These models performed well when applied to just a handful of plant species and illness types. The ability of these models to predict the illnesses of many different plant species has yet to be investigated.

11.4   Proposed Architecture for Prediction of Plant Diseases

The dataset containing images of the plant disease defect to be classified was split into a training and a testing dataset with a test size of 30–20%. In pre-processing, the images are reshaped and resized into appropriate forms to train the model. Since the capability of machine learning to produce characteristics on its own is one of its most important and advantageous characteristics, deep learning pre-processing does not concentrate on feature extraction. Pre-processing is, therefore, primarily concerned with expanding the number of images through data augmentation and scaling the input photos to match the model input parameters.

The following steps are followed in the proposed method as shown in Figure 11.2.

1. Dataset
2. Pre-processing the data
3. Training the models
4. Testing the models
5. Classification of the models

The dataset containing images of the plant disease defect to be classified was split into a training and a testing dataset with a test size of 30–20%. In pre-processing, the images are reshaped and resized into appropriate forms to train the model. Since the capability of machine learning to produce characteristics on its own is one of its most important and advantageous characteristics, deep learning pre-processing does not concentrate on

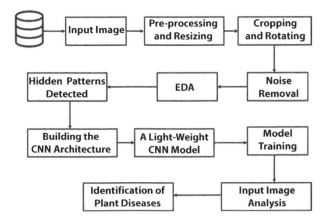

Figure 11.2 Proposed lightweight CNN architecture for prediction of plant diseases.

feature extraction. Pre-processing is, therefore, primarily concerned with expanding the number of images through data augmentation and scaling the input photos to match the model input parameters.

First, we removed the dataset's disruptive, hazy, and out-of-focus images to perform pre-processing. The brightness bending method is used on test images with excessive background merged. A few test images have poor contrast because of the overcast weather and dim lighting, making it challenging to tell the primary grains from a distance. This kind of lack of contrast imagery will result in a decrease in our algorithm's reliability. We employed the Contrast Restricted Dynamic Probabilistic Equalization technique to improve a few of the lower-contrast test images [37]. The input image's distribution reveals that the brightness values are close, making it difficult to differentiate between identical brightness levels. As a result, the input image had minimal contrasts and no discernible details. The intensity values are dispersed across every whole drama in the histogram equalized image, which makes it easier to understand the details. Additionally, the CLAHE cutting approach aids in lowering the noise in the input photos [38].

Augmenting an image was done following pre-processing to boost the number of training images. Our dataset only contains a few images for each class; thus, we used various image augmentation techniques to improve interactivity and account for all potential connections. Various image augmentation techniques have included flipping the input image, skewing, brightness change, randomized movement, spin by multiples of 30 degrees, bruit and noise, shear alteration, and randomized magnification. Ten image enhancements have been created for every sample image using this process. As a result, our dataset is more diverse and considers all potential variants of the sample photos.

To adapt previously acquired knowledge to the problem, reinforcement learning [39] uses that knowledge for a similar activity or domain. Typically, models trained for image classification on sizable datasets are employed and modified for the study's particular dataset. One strategy frequently used to adjust a pre-trained model for various classification objectives is swapping out the last network layers (also known as the dense layers). After that, the model is trained, but only the newly added layers are trainable—all other network layers are frozen during training. The refinement of this strategy is also frequently employed. In addition to training the newly added layers, fine-tuning permits training additional base model layers, usually the network's deeper convolutional layers. Reinforcement learning is typically used when the examined dataset is tiny and there are not enough samples to train a CNN model from scratch. The results of the plant disease defects are classified using the models.

a) Convolutional Neural Network (CNN)

CNN, or convolutional neural network, has become a renowned deep learning algorithm. Most CNN models require images as the inputs and then recognize/classify/predict their features. Convolutional Neural Network processes these images and identifies them based on certain features. Convolutional Neural Network has gained so much popularity in artificial neural networks. This circumstance is because it is used mostly in every field, like in this project, for hand gesture recognition.

b) Convolutional Layer

Multiple convolutional layers in CNN extract low-level features based on whatever you want. To simplify the notion, initial convolutional layers provide lower-level features (like lines and edges), whereas farther convolutional layers give higher-level features. The result of further layers depends on the inputs from lower-level features. This is similar to how vision works in humans.

c) Pooling Layer

The main purpose of CNN is classification in most cases. However, high-dimensional image data are tough to be dealt with. This is why dimensionality reduction is done, and this is done in the pooling layer. Pooling mainly reduces the spatial dimension of image-based mathematical operations:

1. This layer works as a noise suppressant.
2. It makes the data invariant for image classification.
3. It captures several structure-based features of those images without curtailing the finer details.

d) Full Connected Layer

This layer can be imagined as a denser series of convolutional and pooling layers. Mainly, it takes the compressed output of previous layers and fits a basic NN perception in order to classify.

e) Exploratory Data Analysis (EDA)

EDA is a method for visually investigating big datasets. Through visualization, one may comprehend the data's structure and spot trends, outliers, and other things [40]. One needs to be familiarized with the principles involved in EDA to use it efficiently. Knowing variables, their types, scales, distributions, and units falls under this category. The statistical theory underlying this type of study must also be understood. Any data scientist who wants to understand the data must analyze the data using a variety of techniques.

In most cases, data scientists employ both non-graphical and graph-based methods. These strategies comprise (1) correlation-based strategies, (2) regression strategies, (3) clustering strategies, and (4) dimension reduction strategies. Beyond what was said, each of these methods has other applications. As an illustration:

1. Correlation can be used to spot outliers and forecast future results.
2. Regression can be used to forecast future values.
3. Clustering can help identify new racial or ethnic groups.
4. Dimensional reduction can help visualize your data.

The user has to upload an image that needs to be classified. The user views the classified image results.

11.5   Experimental Results and Discussion

The results are simulated using the Python 3.6 software in which Numpy, Flask, TensorFlow, Keras, and OS libraries are used. Using the proposed architecture, the lightweight CNN model's performance for the test image data presented was 78% and 87%. This reinforcement learning Neural

Network model employed a single rate learning policy for all trainable parameters. As learning progresses, the training loss also progressively decreases. It is safe to infer from the corresponding rate that model accuracy is steadily increasing, whereas model loss is steadily decreasing. Eventually, the CNN model created will produce the following total test accuracy percentage, shown in Figures 11.3 and 11.4.

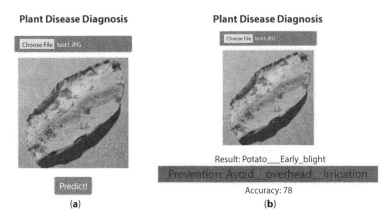

Figure 11.3 Proposed system GUI for plant disease diagnosis. (a) Input image. (b) Output image with an accuracy of 78.

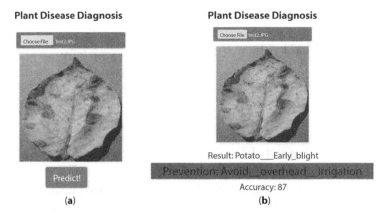

Figure 11.4 Proposed system GUI for plant diseases diagnosis. (a) Input image. (b) Output image with an accuracy of 87.

The proposed system is highly accurate after using the trained and tested model. The system also proved to be effective against real-time data entries. Besides, the pre-processing phase is very time-efficient. Therefore, the algorithm is beneficial for situations where data have to be pre-processed within a short period. The existing systems could be more robust in prediction. They have yet to evolve according to technological advancements and time. The results could be more satisfactory and may lead to erroneous plant disease prediction. The proposed system performs better than all previous ones regarding accuracy and time efficiency.

11.6 Conclusion

Plant diseases reduce agricultural productivity, putting the food supply at risk. As a result, early detection of these illnesses leads to a significant increase in crop output. Using the reinforcement learning model, lightweight CNN architecture was implemented in early prediction to identify plant diseases in different growth stages. Crop diseases can be discovered by examining the structure of leaf images for distortion. Our proposed method improves diagnosis accuracy and is significantly useful to the subsequent strategy. This research will aid in detecting maize crops infected with plant diseases. Designing a lightweight and precise early-detection and alert platform for plant disease illnesses and pest attacks can enable plant conservation in the early stages and increase yields, purity, marketing, and worldwide selling competence.

References

1. Bu, F. and Wang, X., A smart agriculture IoT system based on deep reinforcement learning. *Future Gener. Comput. Syst.*, 99, 500–507, 2019.
2. Zhang, Z., Boubin, J., Stewart, C., Khanal, S., Whole-field reinforcement learning: A fully autonomous aerial scouting method for precision agriculture. *Sensors*, 20, 22, 6585, 2020.
3. Osroosh, Y., Khot, L.R., Peters, R.T., Economical thermal-RGB imaging system for monitoring agricultural crops. *Comput. Electron. Agric.*, 147, 34–43, 2018.
4. Sasikumar, A., Ravi, L., Kotecha, K., Indragandhi, V., Subramaniyaswamy, V., Reconfigurable and hardware efficient adaptive quantization model-based accelerator for binarized neural network. *Comput. Electr. Eng.*, 102, 108302, 2022.

5. Sankar, H., Subramaniyaswamy, V., Vijayakumar, V., Arun Kumar, S., Logesh, R., Umamakeswari, A.J.S.P., Intelligent sentiment analysis approach using edge computing-based deep learning technique. *Softw.: Pract. Exp.*, 50, 5, 645–657, 2020.

6. Rahman, C.R., Arko, P.S., Ali, M.E., Khan, M.A., II, Apon, S.H., Nowrin, F., Wasif, A., Identification and recognition of rice diseases and pests using convolutional neural networks. *Biosyst. Eng.*, 194, 112–120, 2020.

7. Brahimi, M., Mahmoudi, S., Boukhalfa, K., Moussaoui, A., Deep interpretable architecture for plant diseases classification, in: *2019 Signal Processing: Algorithms, Architectures, Arrangements, and Applications (SPA)*, pp. 111–116, 2019.

8. Shaikh, T.A., Rasool, T., Lone, F.R., Towards leveraging the role of machine learning and artificial intelligence in precision agriculture and smart farming. *Comput. Electron. Agric.*, 198, 107119, 2022.

9. Vijayakumar, V., Vairavasundaram, S., Logesh, R., Sivapathi, A., Effective knowledge based recommender system for tailored multiple point of interest recommendation. *Int. J. Web Portals (IJWP)*, 11, 1, 1–18, 2019.

10. Shafi, U., Mumtaz, R., García-Nieto, J., Hassan, S.A., Zaidi, S.A.R., Iqbal, N., Precision agriculture techniques and practices: From considerations to applications. *Sensors*, 19, 17, 3796, 2019.

11. Palomino, W., Morales, G., Huamán, S., Telles, J., PETEFA: Geographic information system for precision agriculture, in: *2018 IEEE XXV International Conference on Electronics, Electrical Engineering and Computing (INTERCON)*, pp. 1–4, 2018.

12. Ravi, L., Devarajan, M., Varadarajan., V., Sangaiah, A.K., Wang, L., Subramaniyaswamy, V., An intelligent location recommender system utilising multi-agent induced cognitive behavioural model. *Enterp. Inf. Syst.*, 15, 10, 1376–1394, 2021.

13. Saleem, M.H., Potgieter, J., Arif, K.M., Plant disease detection and classification by deep learning. *Plants*, 8, 11, 468, 2019.

14. Polese, D., Maiolo, L., Pazzini, L., Fortunato, G., Mattoccia, A., Medaglia, P.G., Wireless sensor networks and flexible electronics as innovative solution for smart greenhouse monitoring in long-term space missions, in: *2019 IEEE 5th International Workshop on Metrology for AeroSpace (MetroAeroSpace)*, pp. 223–227, 2019.

15. Ahmed, N., De, D., Hussain, I., Internet of Things (IoT) for smart precision agriculture and farming in rural areas. *IEEE Internet Things J.*, 5, 6, 4890–4899, 2018.

16. Sasikumar, A., Subramaniyaswamy, V., Jannali, R., Rao, V.S., Ravi, L., Design and area optimization of CMOS operational amplifier circuit using hybrid flower pollination algorithm for IoT end-node devices. *Microprocess. Microsyst.*, 93, 104610, 2022.

17. Chhaya, L., Sharma, P., Kumar, A., Bhagwatikar, G., IoT-based implementation of field area network using smart grid communication infrastructure. *Smart Cities*, 1, 1, 176–189, 2018.

18. Asaithambi, S., Ravi, L., Kotb, H., Milyani, A.H., Azhari, A.A., Nallusamy, S., Subramaniyaswamy, V., An energy-efficient and blockchain-integrated software defined network for the industrial internet of things. *Sensors*, 22, 20, 7917, 2022.

19. Khaki, S. and Wang, L., Crop yield prediction using deep neural networks. *Front. Plant Sci.*, 10, 621, 2019.

20. Sardogan, M., Tuncer, A., Ozen, Y., Plant leaf disease detection and classification based on CNN with LVQ algorithm, in: *2018 3rd International Conference on Computer Science and Engineering (UBMK)*, pp. 382–385, 2018.

21. Rupavatharam, S., Kennepohl, A., Kummer, B., Parimi, V., Automated plant disease diagnosis using innovative android app (plantix) for farmers in Indian state of Andhra Pradesh. *Phytopathol. (TSI)*, 108, 10, 2018. http://oar.icrisat.org/id/eprint/11014.

22. Iniyan, S. and Jebakumar, R., Mutual information feature selection (MIFS) based crop yield prediction on corn and soybean crops using multilayer stacked ensemble regression (MSER). *Wirel. Pers. Commun.*, 126, 3, 1935–1964, 2022.

23. Nandhini, S. and Ashokkumar, K., An automatic plant leaf disease identification using densenet-121 architecture with a mutation-based henry gas solubility optimization algorithm. *Neural Comput. Appl.*, 34, 1–22, 2022.

24. Sasikumar, A., Ravi, L., Kotecha, K., Saini, J.R., Varadarajan, V., Subramaniyaswamy, V., Sustainable smart industry: A secure and energy efficient consensus mechanism for artificial intelligence enabled industrial internet of things. *Comput. Intell. Neurosci.*, 2022, 1–12, 2022.

25. Selvan, N.S., Vairavasundaram, S., Ravi, L., Fuzzy ontology-based personalized recommendation for internet of medical things with linked open data. *J. Intell. Fuzzy Syst.*, 36, 5, 4065–4075, 2019.

26. Oh, H.M., Lee, H., Kim, M.Y., Comparing convolutional neural network (CNN) models for machine learning-based drone and bird classification of anti-drone system, in: *2019 19th International Conference on Control, Automation and Systems (ICCAS)*, pp. 87–90, 2019.

27. Zhu, L., Li, Z., Li, C., Wu, J., Yue, J., High performance vegetable classification from images based on alexnet deep learning model. *Int. J. Agric. Biol. Eng.*, 11, 4, 217–223, 2018.

28. Zhu, N., Liu, X., Liu, Z., Hu, K., Wang, Y., Tan, J., Guo, Y., Deep learning for smart agriculture: Concepts, tools, applications, and opportunities. *Int. J. Agric. Biol. Eng.*, 11, 4, 32–44, 2018.

29. Shah, A., Dubey, A., Hemnani, V., Gala, D., Kalbande, D.R., Smart farming system: Crop yield prediction using regression techniques, in: *Proceedings of*

International Conference on Wireless Communication: ICWiCom, pp. 49–56, 2017.

30. Logesh, R., Subramaniyaswamy, V., Malathi, D., Senthilselvan, N., Sasikumar, A., Saravanan, P., Manikandan, G., Dynamic particle swarm optimization for personalized recommender system based on electroencephalography feedback. *Biomed. Res.*, 28, 13, 5646–5650, 2017.

31. Agarwal, M., Bohat, V.K., Ansari, M.D., Sinha, A., Gupta, S.K., Garg, D., A convolution neural network based approach to detect the disease in corn crop, in: *2019 IEEE 9th International Conference on Advanced Computing (IACC)*, pp. 176–181, 2019.

32. Ramamurthy, K., M.H., Anand, S., Mathialagan, P.M., Johnson, A., R.M., Attention embedded residual CNN for disease detection in tomato leaves. *Appl. Soft Comput.*, 86, 105933, 2020.

33. Ma, N., Zhang, X., Zheng, H.T., Sun,J., Shufflenet v2: Practical guidelines for efficient cnn architecture design, in: *Proceedings of the European Conference on Computer Vision (ECCV)*, pp. 116–131, 2018.

34. Islam, F., Hoq, M.N., Rahman, C.M., Application of transfer learning to detect potato disease from leaf image, in: *2019 IEEE International Conference on Robotics, Automation, Artificial-Intelligence and Internet-of-Things (RAAICON)*, pp. 127–130, 2019.

35. Akila, M. and Deepan, P., Detection and classification of plant leaf diseases by using deep learning algorithm. *Int. J. Eng. Res. Technol. (IJERT)*, 6, 7, 1–5, 2018.

36. Chen, J., Wang, W., Zhang, D., Zeb, A., Nanehkaran, Y.A., Attention embedded lightweight network for maize disease recognition. *Plant Pathol.*, 70, 3, 630– 642, 2021.

37. Anuranjeeta, Sharma, S., Sharma, N., Singh, M., Shukla, K.K., Enhancement and segmentation of histopathological images of cancer using dynamic stochastic resonance. *Int. J. Med. Eng. Inform.*, 12, 2, 180–193, 2020.

38. Setiawan, A.W., Mengko, T.R., Santoso, O.S., Suksmono, A.B., Color retinal image enhancement using CLAHE, in: *International Conference on ICT for Smart Society*, pp. 1–3, 2013.

39. Gandhi, R., Deep reinforcement learning for agriculture: principles and use cases, in: *Data Science in Agriculture and Natural Resource Management*, pp. 75–94, 2022.

40. Kumar, M., Kumar, A., Palaparthy, V.S., Soil sensors-based prediction system for plant diseases using exploratory data analysis and machine learning. *IEEE Sensors J.*, 21, 16, 17455–17468, 2020.

12

Investigation of Feature Fusioned Dictionary Learning Model for Accurate Brain Tumor Classification

P. Saravanan[1], V. Indragandhi[2], R. Elakkiya[3] and V. Subramaniyaswamy[4]*

[1]*School of Computer Science and Engineering, Vellore Institute of Technology, Chennai, India*
[2]*School of Electrical Engineering, Vellore Institute of Technology, Vellore, India*
[3]*Department of Computer Science, BITS Pilani, Dubai Campus, Dubai, United Arab Emirates*
[4]*School of Computing, SASTRA Deemed University, Thanjavur, India*

Abstract

Brain tumors, which are one of the most prevalent and deadliest diseases, have an extremely low life expectancy when they are at their most advanced stage. The fate of brain tumors may be significantly influenced by early diagnosis. Brain tumors are diagnosed from MRI images using standard symptoms. However, using an MRI to determine the type of tumor is time-consuming, challenging, and error-prone as the number of cases increase. Given the significant anatomical and geographical heterogeneity of the brain tumor's surrounding area, automatically classifying brain tumors is a highly difficult process. In this work, we proposed a deep learning approach that considers the hybridization of CNN and LSTM with Dictionary Learning. Further, the proposed model is optimized using Feature Fusion. The experimental results prove that the proposed Feature Fusioned Dictionary Learning Model outperforms in terms of accuracy and improves the classification process by 4.58% from the core model.

Keywords: Brain tumor classification, convolutional neural network, dictionary learning, feature fusion, ResNet-50, long short-term memory network

Corresponding author: swamy@cse.sastra.ac.in

Elakkiya, R. and Subramaniyaswamy V. (eds.) Cognitive Analytics and Reinforcement Learning: Theories, Techniques and Applications, (223–238) © 2024 Scrivener Publishing LLC

12.1   Introduction

Brain tumors are abnormal cell growths in the brain or the tissues around the brain. They may be benign or malignant, a cancerous condition. Any area of the brain, including the cerebral hemispheres, brainstem, cerebellum, or meninges (the brain's covering), can develop a brain tumor. There is still much to learn about the precise etiology of brain tumors. There are, however, certain known risk factors, such as ionizing radiation exposure, a few genetic abnormalities, a family history of brain tumors, and immune system issues. Additionally, several studies raise the possibility that using a cell phone may increase the risk of developing a brain tumor. Depending on a brain tumor's size, location, and rate of growth, several symptoms may be present [1]. Headaches, seizures, altered vision or hearing, trouble speaking or comprehending language, limb weakness or numbness, balance issues, and personality changes are among the common symptoms. A healthcare expert should be consulted for a precise diagnosis because these symptoms can also be brought on by other illnesses.

According to the American Society of Clinical Oncology (ASCO) [2], a tumor that originates in the brain or spinal cord is referred to as a primary tumor. A primary malignant tumor of the brain or spinal cord will be identified in 24,810 individuals (14,280 men and 10,530 women) in the US in 2023. Less than 1% of people will experience this form of tumor in their lives; 85% to 90% of all primary Central Nervous System (CNS) malignancies are brain tumors. In addition, 5,230 kids under the age of 20 are anticipated to receive a CNS tumor diagnosis in the US in 2023. The 10th biggest cause of mortality for women and men is cancer of the neurological system, including the brain. In 2023, it is predicted that primary malignant brain and CNS tumors would be the leading cause of 18,990 deaths (11,020 males and 7,970 females). According to the National Library of Science [3], brain tumors have recently become one of the top 10 tumors impacting the Indian population. Up to 10 out of every 100,000 people in India are diagnosed to have central nervous system tumors, with 2% of those tumors being malignant.

12.1.1   Importance of Accurate and Early Diagnosis and Treatment

Depending on the kind, grade, location, and other circumstances, there are vast variations in the survival rates of brain tumors. Depending on the exact tumor type and grade, the 5-year survival rate for malignant brain tumors ranges from around 15% to 40%. The prognosis for benign brain tumors is often better since they are more likely to be completely removed and have better long-term survival rates. The risk to life can be decreased with an early and precise diagnosis. Here are numerous strong reasons why timely diagnosis and treatment are crucial.

- *Proper Disease Management:* Knowing the precise condition that an individual is suffering from requires an accurate diagnosis. It makes it possible for medical experts to create a treatment plan that is suitable and adapted to the patient's requirements.
- *Treatment Selection and Effectiveness:* Healthcare professionals can select the best course of treatment for patients with the help of an accurate diagnosis. Treatment options for various forms of brain tumors may include surgery, radiation therapy, chemotherapy, or targeted therapy. The best possible therapies are administered to patients with the help of a precise diagnosis, increasing their chances of a positive outcome and improving their quality of life.
- *Precision Medicine and Personalized Therapies:* Precision medicine, which strives to give focused medicines based on a patient's specific disease conditions, starts with an accurate diagnosis. Medical practitioners may find genetic or molecular markers that relate to the tumor and modify treatment plans using accurate diagnostic data.

Underdiagnosis of brain tumors is also risky since it lowers the effectiveness of treatments and the chance of survival. Patients who have correct diagnoses obtain effective treatments and live a long life.

Brain tumors are diagnosed using Magnetic Resonance Imaging (MRI), which is one of the most used high-precision approaches for identifying and diagnosing cancer [4]. However, using an MRI to determine the kind of tumor is time-consuming, challenging, and error-prone, necessitating the expertise of highly qualified radiologists. The variety of tumors makes it occasionally impossible to detect characteristics in MRI scans that would allow for accurate and early decision-making. As a result, manual diagnosis is difficult for humans to rely on. As a result, using machine learning and deep learning techniques to diagnose medical imagery like MRI imaging has become essential.

12.1.2   Role of Machine Learning in Brain Tumor Classification

Machine learning is essential in classifying brain tumors since it brings about several important advantages and advances. The following are some of the main factors that make deep learning important in classifying brain tumors:

- The accuracy of brain tumor categorization may be improved using machine learning methods. Machine learning algorithms can recognize complicated patterns and subtle traits that may be challenging for human specialists to notice by evaluating vast amounts of data, including medical imaging, genetic data, and clinical features. This might improve the categorization of brain cancers, improve diagnostic precision, and help guide therapy choices.
- Machine learning algorithms can quickly process and evaluate massive amounts of data, which improves the effectiveness of the categorization process. This may facilitate quicker decision-making and perhaps improve patient outcomes by reducing the amount of time needed for diagnostic and treatment planning.
- Machine learning algorithms are capable of combining and interpreting data from several multimodal sources, such as genetics, clinical data, and medical imaging. This enables full and in-depth analysis of brain tumors by taking into account a number of informative factors.
- Machine learning algorithms enable the creation of personalized medical procedures for classifying brain cancers. By considering individual patient factors such as demographic data, imaging results, genetic profiles, and treatment histories, deep learning models may adapt categorization and treatment methods to each patient's particular needs. This unique approach might improve patient care and treatment outcomes.

All machine learning models for brain tumor classification use Computer-Aided Diagnosis (CAD) on MRI images. However, the image datasets used have sparsity features that may lead to poor performance of the model.

12.1.3   Sparsity Issues in Brain Image Analysis

The majority of the values for features with sparse data are zero. In learning models, sparse features can lead to issues including overfitting and poor outcomes. When creating models, it is essential to comprehend why this occurs. To solve problems brought on by sparse features, a variety of techniques are available, including dictionary learning and dimensionality reduction [5]. The space and temporal complexity of the model will grow if it has a lot of sparse features. Tree-based models will have higher depth to account for all characteristics, whereas linear regression models will fit more coefficients. If the features have limited data, model algorithms and diagnostic measures may react in unknown ways. When the data are sparse, Kuss [6] demonstrates that goodness-of-fit tests are invalid.

Models fit the training data noise if there are too many features. Overfitting is the term for this. When models are used in production, over-fitting prevents them from generalizing to new data. This has a detrimental effect on a model's ability to forecast. Even while sparse features could have predictive potential, certain models might undervalue their significance and favor richer characteristics. This is a common behavior of tree-based models. Random forests, for instance, overestimate the value of characteristics with more categories compared to features with fewer categories. Considering all the above issues and understanding the importance of early and accurate diagnosis of brain tumors in this study, we proposed a feature fusioned dictionary learning model for brain tumor classification with reduced sparsity in brain image analysis. This chapter is further organized as follows. Section 12.2 discusses various existing works on brain tumor identification using traditional image processing, Computer-Aided Diagnosis, and Machine Learning models. Section 12.3 explains the architectural representation of the proposed model and its steps. Section 12.4 describes the implementation setup and discusses the experimental results. Finally, the conclusion of the work and future improvements are given in section 12.5.

12.2 Literature Review

Deepak [7] has proposed a classification method that employs a pre-trained GoogLeNet to extract features from brain MRI images and implements the idea of deep transfer learning. The collected characteristics are classified using integrated, tested classifier models. On an MRI dataset from Figshare, the experiment uses a fivefold cross-validation approach at the patient level. The suggested approach outperforms all cutting-edge techniques with a mean classification accuracy of 98%. The study also employed the area under the curve (AUC), precision, recall, F-score, and specificity as performance indicators. By testing the system with fewer training examples, the chapter also solves a practical issue. The study's findings suggest that transfer learning is a practical method when there are few medical pictures available.

A hybrid classification model has been proposed in [8], which integrates Discrete Wavelet Transform (DWT), a genetic algorithm, and a Support Vector Machine. DWT extracts the features, the genetic algorithm reduces the number of features, and SVM classifies the brain tumors. The proposed system has been evaluated using the SICAS Medical Image dataset with classes of brain tumors. The authors have used the MATLAB 2015a platform for implementation and kurtosis, entropy, smoothness, root mean square error (RMSE), and correlation have been used as performance measures. The simulation results reveal that the hybrid strategy gives higher performance by increasing accuracy and lowering the RMS error.

In the research article [9], an automated classifier to distinguish between brain MRI scans with cancer and those without malignancy has been introduced. Considering form, texture, and intensity, a feature set is selected for each applicant's wound. The Support Vector Machine (SVM) classifier is then used on the collection of features to compare the proposed framework's precision using various cross-validations. Three benchmark datasets, including Harvard, RIDER, and Local, are used to verify the suggested technique. The proposed system has produced an average of 97.1% accuracy. Also, the area under the curve was 0.98, the sensitivity was 91.9%, and the specificity was 98.0% for the procedure. It is proved that compared to current techniques, this model can be utilized to detect the tumor more precisely and with less processing time.

Sharif *et al.* [10] proved that the fusion of texture and geometrical features has improved the accuracy in the early detection of brain tumors. The proposed system has performed well by combining Fast Non-Local Mean (FNLM), Ostu algorithm feature fusion, and classification using multiple classifiers. FNLM is used for enhancement, the segmentation is taken care of by Ostu algorithm.

The segmentation outcomes are enhanced by using conditional random fields to the Deep Discriminative Model (DDM) [11]. The set of pre-P-norm filtering and thresholding convolutional operations may be learned using gradient descent training with a special geometric P-norm-based loss function. This latter element will be optimized by the RANSAC layer, which is intended to increase the actual model's strength in comparison to the strongest false model. The precision and reliability are tested using MRI sequences from the Multimodal Brain Tumor Image Segmentation Challenge (T1, T2, T2c, and FLAIR). The trials show that our DPGM-DDM architecture produces high-quality segmentation results, and the average Dice is as high as 98.4% for enhancing tumor areas.

In [12], the authors suggested a non-invasive CAD system based on brain MRIs that can help radiologists and doctors identify not only the existence of glioma tumors but also their specific class. It performs three stages of classification using multiple segmentation techniques for various pulse sequences, the fusion of texture features, and an ensemble classifier. When a tumor is found at the first level of categorization, the tentorium of the brain is used to examine its position, and the tumor is then classed as supratentorial or infratentorial at the second level. At the third level of categorization, the method determines the tumor type based on the morphological and inherent properties of the tumor (area, perimeter, firmness, and orientation).

To segment and identify brain tumors, a feature selection strategy based on active deep learning [13] is proposed. The Inception V3, which is a pre-trained CNN model, is used for extracting deep features. The extracted features are further concatenated with Dominant Rotated Local Binary Patterns (DRLBP) and a softmax classifier is applied. The result is optimized by particle swarm optimization (PSO). BRATS2017 and BRATS2018 datasets are used in the segmentation and classification phases of the trials. The suggested model generated results with an average accuracy of above 92%. The proposed system has outperformed well in the segmentation and classification of brain tumor images.

Spatiospatial models [14] can learn certain spatial and temporal interactions while using less processing power. The proposed method classifies several types of brain tumors using ResNet (2+1)D and ResNet Mixed Convolution. It was shown that both of these models outperformed ResNet18, a model that only used 3D convolutions. It was also demonstrated that performance is improved by pre-training the models on a separate, even unrelated dataset before training them for the goal of cancer classification. Pre-trained ResNet Mixed Convolution was shown to be the most effective model in these trials. The system has achieved 96.98% test accuracy. The existing research approaches for brain tumor classification have been focused on different machine learning and deep learning combinations and variants that aim to improve the MRI image segmentation analysis. However, this proposed system aims at improving the performance of tumor classification by reducing the sparsity.

12.3   Proposed Feature Fusioned Dictionary Learning Model

This section describes the methodologies, algorithms, and steps of the suggested approach that are utilized to classify MRI brain imaging data.

An efficient automatic classification mechanism called Feature Fusioned Dictionary Learning Model (FFDLM) is proposed, which extracts discriminative features by integrating CNN with Dictionary Learning. This can be further improved by replacing CNN with a feature fusion model (ResNet50 and CNN+LSTM). The advantages of using this technique are as follows:

- It learns feature representation using multi-layered Dictionary Learning
- Encoding in non-linear space
- Encoding using a multi-layered Dictionary means multiple projections

The proposed model consists of six distinct processes, the first of which involves the collection of the dataset. The second phase involves data pre-processing, which includes operations like thresholding and histogram pre-processing. Then dictionary learning is applied to reduce the sparsity in the dataset. Feature extraction is applied in the fourth phase, and features are fusioned further. Finally, the experiment to find the brain tumor is done in the sixth step using the CNN-based ResNet model and CNN-LSTM hybrid learning approach. Figure 12.1 shows the overall architecture of the proposed system.

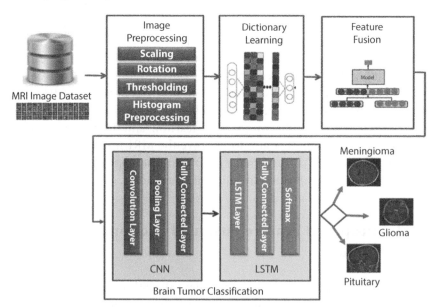

Figure 12.1 Architecture of proposed FFDLM for brain tumor classification.

The dataset is collected from Kaggle-Figshare, which has MRI images of three different types of brain tumors. The dataset consists of 708 images of meningioma, 1426 images of glioma, and 930 slices of pituitary. The images are pre-processed with basic pre-processing techniques such as for improvement. Further histogram pre-processing has been applied. A gray-scale value distribution known as an image histogram displays the frequency of occurrence of each gray-level value. The abscissa spans from 0 to 255 for an image size of $1024 \times 1024 \times 8$ bits, and the number of pixels is calculated by 1024×1024. Procedures for improving images sometimes include altering the original histograms.

Once the images are pre-processed, the feature selection process is done based on dictionary learning. For feature extraction and data representation, dictionary learning [15] is a method used in deep learning and machine learning, and signal processing. From the input data, a dictionary of representative atoms or basic functions is learned. The representation of each data point is thus a sparse linear mixture of these dictionary components. Dictionary learning is trained using an optimization problem with constraints, where the optimization is done to minimize the difference between multiplying the learned matrix "D" and its coefficient matrix "R" from the input data matrix "X". The goal is to reduce the reconstruction error while promoting coefficient sparsity.

Dictionary learning is advantageous for applications like image denoising and super-resolution because it provides localized and interpretable features that are directly learned from the input. Although deep learning models may produce results that are comparable from every phase, dictionary learning is still useful for some applications and acts as a preprocessing step in deep learning pipelines.

Feature fusion aids in the complete learning of image features for the description of their rich internal information. The process of combining the many extracted features from several databases into a single feature file is known as feature fusion [16]. To improve efficiency and shorten the system's execution time, the major objective of fusion technology in brain tumor diagnostics is to merge discriminative and trustworthy data from all derived characteristics into one vector. Reduced feature size and removal of noisy features are the main goals of feature fusion.

A convolutional neural network (CNN) is a form of artificial neural network that specializes in learning technique component knowledge and is used for image recognition. Typically employing machine vision, which contains image and video recognition, along with recommender systems and language communication process (NLP), CNN is a sophisticated image processing, computing technology that uses deep learning to do both generative and descriptive tasks. A CNN employs a method like a multilayer viewpoint that has been created for less process requirements. A system that is significantly more effective and simpler in training data for both language communication and image processing results from the elimination of restrictions and improvement in image processing potency.

When multiple types of networks are combined to create hybrid models, one of the most exciting and practically valuable neural models will be designed [17]. Because they enable us to condense features for prediction, LSTM networks are certainly the most successful RNNs. Because they are unprepared to handle such inputs, we cannot directly input the RGB picture tensor. The typical LSTM cannot be used to model input having a spatial structure, such as pictures. To extract features from the image, we may utilize the deep CNN architecture. These features can then be input into the LSTM architecture to produce the caption. The system has used the CNN model with the method of probability factorization and suggested the Convolutional MF quality service guideline.

12.4   Experimental Results and Discussion

To validate the proposed FFDLM, the brain tumor dataset from Figshare has been used. The dataset consists of 3064 MRI images that are T1-weighted and contrast-enhanced. They are collected from 233 brain tumor patients and normal persons. It consists of three classes of brain tumors called pituitary, meningioma, and glioma. The proposed system is implemented in Python with Jupyter execution environment. the calculated accuracy is compared to all other state-of-the-art methodologies.

The suggested brain tumor classification scheme's effectiveness is determined by calculating the training accuracy, validation accuracy, and validation loss. Training accuracy refers to a model's accuracy on the data it was trained on. The validation accuracy represents the performance of the proposed model on new data. Figure 12.2 shows the relationship between the training accuracy and validation accuracy of our model with different epochs. We considered 0 to 9 epochs, and up to 7 epochs, the training accuracy is lower than the validation accuracy, and at the eighth epoch, it is very close to validation accuracy.

Loss functions are used to further assess the model. The training loss is a measure of how well a deep learning model matches the training data. It is calculated using the sum of errors for each sample in the training set.

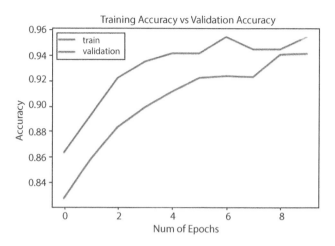

Figure 12.2 Performance of proposed model—training accuracy vs. validation accuracy.

Additionally, it is critical to keep in mind that every epoch yields a measurement of the training loss. On the other hand, a deep-learning model's validation loss statistic measures how well it performs on the validation set. A portion of the dataset designated to test the model's performance is known as the validation set. The validation loss is determined from the total of the mistakes for each sample in the validation set and is comparable to the training loss. A comparison of Training loss and Validation of our model has been shown in Figure 12.3. The model is executed from 0 to 9 epochs. Initially, the training loss was 44%, and the validation loss was 36%. The system is improved by increasing the epochs. The system has obtained minimum training and validation losses in the ninth epoch.

The overall proposed system performance is evaluated by calculating the accuracy value with multiple epochs and the average accuracy of the system considered for the comparison. For comparison, we have considered different variants of the traditional classification models. The performance evaluation has been done in three phases. In the first phase, the selected basic classification models (1) Convolutional Neural Network (CNN), (2) Support Vector Machine (SVM), and (3) Random Forest (RF) are implemented directly and the average accuracies are measured. In phase two, the selected classification models are implemented with Dictionary Learning and referred to as (4) CNN-DL, (5) SVM-DL, and (6) RF-DL, respectively. In the third phase, the ResNet-50 with Dictionary Learning (RESTNET-DL), the hybrid version of CNN and LSTM with Dictionary Learning (CNN+LSTM-DL), and its feature fused model (Feature Fused DL) have been implemented. The average accuracies of

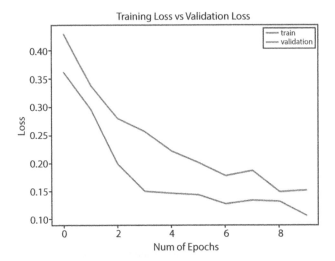

Figure 12.3 Performance of proposed model—training loss vs. validation loss.

the systems have been measured and compared in Table 12.1 and Figure 12.4. From this visualization, it is evident that the core CNN, SVM, and RF produce 90.86%, 88.58%, and 83.5% of average accuracy. When Dictionary

Table 12.1 Average accuracy produced by various classification models.

S. no.	Models	Average accuracy (%)
1	Convolutional Neural Network (CNN)	90.86
2	Support Vector Machine (SVM)	88.58
3	Random Forest (RF)	83.5
4	CNN with Dictionary Learning (CNN-DL)	94.19
5	SVM with Dictionary Learning (SVM-DL)	90.85
6	RF with Dictionary Learning (RF-DL)	86.6
7	ResNet-DL	93.16
8	CNN+LSTM-DL	94.33
9	Proposed FEATURE FUSION DL(FFDLM)	95.44

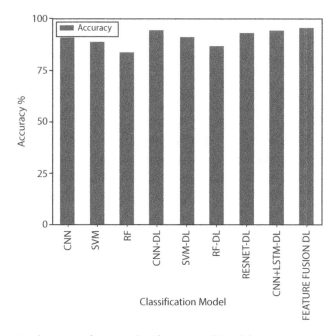

Figure 12.4 Performance of various classification model and their accuracy.

Learning is added, they have improved to 94.19%, 90.85%, and 86.6%, respectively.

When ResNet is implemented with Dictionary Learning, the average accuracy is 93.16%. Finally, the proposed system without feature fusion has given 94.33% of average accuracy and with feature fusion, the average accuracy is 95.44%. This is the highest accuracy among all other accuracies implemented in this work. Our proposed Feature Fusioned Dictionary Learning Model (FFDLM) outperforms brain tumor classification in classifying the given MRI images.

12.5   Conclusion and Future Work

In this research, we proposed a deep learning model called Feature Fusioned Dictionary Model (FFDLM) for brain tumor classification from the MRI images. The model aims for early identification of three classes of brain tumors called pituitary, meningioma, and glioma. A Python implementation has been done for different deep-learning scenarios. The selected core classification models are CNN, SVM, and RF. These core models are further tuned as CNN-DL, SVM-DL, and RF-DL by adding Dictionary

Learning logic to reduce the sparsity. To validate the proposed FFDLM, the brain tumor dataset from Figshare with 3064 MRI images has been used. The above-proposed models have given 1.87% average improved accuracy when Dictionary Learning is added. Further, when combining CNN and LSTM along with Dictionary Learning, the accuracy improvement is 3.47 and the same when feature fusion is 4.58%. From the results, it is concluded that the proposed hybrid model with Dictionary Learning and feature fusion called FFDLM outperforms brain tumor classification. In the future, it is planned to add an Application Programming Interface that directly takes MRI images from an MRI scanner, predicts the result, and displays the result with multiple information.

References

1. https://braintumor.org/brain-tumors/about-brain-tumors/brain-tumor-facts/
2. https://www.cancer.net/cancer-types/brain-tumor/statistics
3. https://timesofindia.indiatimes.com/life-style/health-fitness/health-news/doctor-lists-5-possible-causes-of-brain-tumour/photostory/100497382.cms?picid= 100497407
4. Badza, M.M. and Barjaktarovic, M.C., Classification of brain tumors from MRI images using a convolutional neural network. *Appl. Sci.*, 10, 6, 1999, 2020.
5. https://www.kdnuggets.com/2021/01/sparse-features-machine-learning-html#:~:text=Sparse%20features%20are%20common%20in,feature%20importances%2C%20and%20high%20variance
6. Kuss, O., Global goodness-of-fit tests in logistic regression with sparse data. *Stat. Med.*, 21, 24, 3789–3801, 2002.
7. Deepak, S. and Ameer, P.M., Brain tumor classification using deep CNN features via transfer learning. *Comput. Biol. Med.*, 111, 103345, 2019.
8. Kumar, S., Dabas, C., Godara, S., Classification of brain MRI tumor images: A hybrid approach. *Proc. Comput. Sci.*, 122, 510–517, 2017.
9. Amin, J., Sharif, M., Yasmin, M., Fernandes, S.L., A distinctive approach in brain tumor detection and classification using MRI. *Pattern Recognit. Lett.*, 139, 118–127, 2020.
10. Sharif, M., Amin, J., Nisar, M.W., Anjum, M.A., Muhammad, N., Shad, S.A., A unified patch based method for brain tumor detection using features fusion. *Cogn. Syst. Res.*, 59, 273–286, 2020.
11. Takrouni, W. and Douik, A., Improving geometric P-norm-based glioma segmentation through deep convolutional autoencoder encapsulation. *Biomed. Signal Process. Control*, 71, 103232, 2022.

12. Gupta, N., Bhatele, P., Khanna, P., Glioma detection on brain MRIs using texture and morphological features with ensemble learning. *Biomed. Signal Process. Control*, 47, 115–125, 2019.
13. Sharif, M., II, Li, J.P., Khan, M.A., Saleem, M.A., Active deep neural network features selection for segmentation and recognition of brain tumors using MRI images. *Pattern Recognit. Lett.*, 129, 181–189, 2020.
14. Chatterjee, S., Nizamani, F.A., Nurnberger, A., Speck, O., Classification of brain tumours in MR images using deep spatiospatial models. *Sci. Rep.*, 12, 1, 1505, 2022.
15. Gu, X., Shen, Z., Xue, J., Fan, Y., Ni, T., Brain tumor MR image classification using convolutional dictionary learning with local constraint. *Front. Neurosci.*, 15, 679847, 2021.
16. Manavalan, R., Automatic identification of diseases in grains crops through computational approaches: A review. *Comput. Electron. Agric.*, 178, 105802, 2020.
17. Alsubai, S., Khan, H.U., Alqahtani, A., Sha, M., Abbas, S., Mohammad, U.G., Ensemble deep learning for brain tumor detection. *Front. Comput. Neurosci.*, 16, 1005617, 2022.

13

Cognitive Analytics-Based Diagnostic Solutions in Healthcare Infrastructure

Akshay Bhuvaneswari Ramakrishnan[1], T. S. Murugesh[2]*, Sini Raj Pulari[3] and Shriram K. Vasudevan[4]

[1]Department of Computer Science and Engineering, SASTRA University, Tamil Nadu, India
[2]Department of Electronics and Communication Engineering, Government College of Engineering Srirangam, Tiruchirappalli, Tamil Nadu, India (On Deputation from Annamalai University, Department of Electronics and Instrumentation Engineering, Faculty of Engineering & Technology, Tamil Nadu, India)
[3]Tutor – Bahrain Polytechnic, ISA Town, Bahrain
[4]Lead Technical – Evangelist (Asia Pacific and Japan), Intel India Pvt. Ltd. Bengaluru, Karnataka, India

Abstract

Cognitive computing, a subfield of artificial intelligence, has the potential to completely transform smart cities and healthcare, which is discussed in this chapter, demonstrating how AI-powered technologies are transforming urban environments and medical practices. By incorporating cognitive systems into the infrastructure of smart cities, resource allocation can be optimized, transportation networks can be improved, and citizen services can be enhanced. Successful cognitive technology implementations in cities such as Singapore and Barcelona are illustrated by case studies, easing the way for smarter urban living. In addition, cognitive computing provides healthcare providers with data-driven insights for accurate diagnosis, tailored therapies, and real-time monitoring in the healthcare industry. Exemplary examples of AI-powered telemedicine platforms illustrate how remote care and virtual meetings enhance access to healthcare and patient outcomes. To realize the full potential of cognitive computing in these crucial domains, however, ethical issues and responsible execution are necessary.

*Corresponding author: tsmurugesh@gmail.com

Elakkiya, R. and Subramaniyaswamy V. (eds.) Cognitive Analytics and Reinforcement Learning: Theories, Techniques and Applications, (239–252) © 2024 Scrivener Publishing LLC

Keywords: Cognitive computing, artificial intelligence, smart cities, healthcare, AI-powered technologies, urban living

13.1   Introduction

Convergence of cognitive computing and artificial intelligence (AI) has emerged as a revolutionary force in the fast-paced and interconnected world of the 21st century, promising groundbreaking improvements in two crucial domains: smart cities and healthcare. This convergence has the potential to revolutionize both of these fields. The development of cognitive computing, which is being driven by cutting-edge artificial intelligence technology, has the potential to transform medical procedures and reimagine what it means to live in a city, thereby moving society toward a future that is more intelligent, more environmentally friendly, and healthier. The idea of smart cities, which represents the unobtrusive incorporation of data-driven technology into urban environments, has sparked the interest of city planners and inventors all over the world. The concept of highly intelligent cities that are able to adapt in a dynamic manner to the requirements of their citizens has emerged from the domain of science fiction and into the realm of a potentially attainable reality thanks to the development of cognitive computing [1]. The capability of cognitive systems to learn from large volumes of data, reason through complicated scenarios, and grasp natural language enables cities to optimize the allocation of resources, streamline transportation networks, and increase citizen services with an efficiency that is unmatched. In the meantime, the healthcare industry is struggling with a wide variety of obstacles, such as the requirement for accurate diagnoses, the requirement for individualized treatment programs, and the desire for resource allocation that is as efficient as possible.

Cognitive computing emerges as a strong ally in this field, as it provides data-driven insights to medical professionals and enhances their ability to make decisions. Not only does the potential for diagnostics powered by artificial intelligence (AI), real-time patient monitoring, and predictive analytics improve patient outcomes, but it also affects the way healthcare services are offered and enjoyed. The objective of this work is to investigate the mutually beneficial link that exists between cognitive computing and the two most important industries, namely, healthcare and smart cities. By analyzing case studies from the actual world, we hope to demonstrate how advances in cognitive technology alter the urban landscape and bring about revolutionary changes in medical procedures. We will demonstrate, by means of thorough analysis, the tangible benefits, the evolving problems, and the ethical considerations that underline the right integration of cognitive computing into these key aspects of the current society.

As we set out on this adventure into the world of cognitive computing in smart cities and healthcare, our primary objective is to shed light on the enormous potential that these game-changing technologies have. The potential of artificial intelligence can be used in a responsible manner, which holds the key to unlocking a future in which cities flourish on intelligence and healthcare becomes truly patient-centric and precise. In order to successfully navigate this transformative terrain, we need to address the complexities of technology, the nuances of urban life, and the fundamental principles of healthcare [2], all while working toward the common goal of making the world not only more intelligent and healthier, but also more egalitarian and morally sound.

13.2   Cognitive Computing in Action

Cognitive computing is a significant step forward for artificial intelligence since it breaks away from the traditional rule-based systems that have been used up until now and enters the world of learning and comprehension more like that of humans. The fundamental goal of cognitive computing is to create computer systems that can reason their way through difficult issues, learn from large amounts of data, and understand natural language. This will allow computers to learn from and reason with data collected from humans, as well as learn from and interpret natural language. This section of the chapter looks into the core principles of cognitive computing and its integration within the larger AI environment [3]. Specifically, it examines how cognitive computing fits into the overall AI ecosystem. The whole system is also depicted as a flowchart in Figure 13.1.

13.2.1   Natural Language Processing (NLP)

The capability of cognitive computing to comprehend and make sense of human language is one of the defining characteristics of this type of computing. Natural Language Processing (NLP) is a type of computer technology that enables machines to comprehend, generate, and respond in human language. This technology helps bridge the communication gap between humans and machines. We investigate the intricacies of natural language processing (NLP), including sentiment analysis, speech

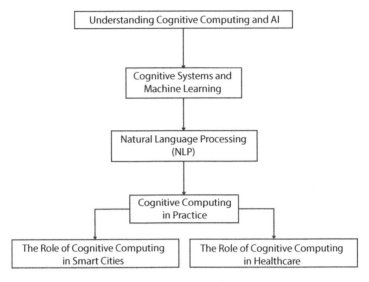

Figure 13.1 Understanding cognitive computing and AI.

recognition, and language translation, with the goal of highlighting the role that these technologies play in making cognitive systems more user-friendly and approachable.

13.2.2   Application of Cognitive Computing in Everyday Life

In order to demonstrate how cognitive computing may be put to use in the real world, we will present some examples of how this technology is now being utilized in a variety of business sectors. The way in which we engage with technology and take in information is being revolutionized by cognitive computing. This can be seen in the rise of virtual assistants such as Siri and Alexa, as well as in the personalized content recommendations made available on streaming platforms. These use cases illustrate the myriad ways in which cognitive computing can have an effect on our day-to-day activities.

13.2.3   The Importance of Cognitive Computing in the Development of Smart Cities

As we learn more about the capabilities of cognitive computing, we are beginning to investigate how it may be used within the framework of smart cities. It is possible for urban environments to become more adaptable, responsive, and efficient by merging cognitive technology with the massive network of interconnected devices and sensors that is present in smart cities [4]. We look at how cognitive computing may improve data analysis, make predictive maintenance easier, and maximize energy efficiency, thereby laying the framework for a more environmentally friendly and livable urban future.

13.2.4 The Importance of Cognitive Computing in the Healthcare Industry

The application of cognitive computing has the potential to completely revolutionize the healthcare industry. The medical community now has access to cognitive systems that can evaluate patient data, provide assistance with difficult diagnoses, and help personalize treatment programs to the specific requirements of individual patients. We study how artificial intelligence-powered cognitive technologies are already reshaping the healthcare industry by enhancing the quality of medical outcomes, simplifying operations, and giving patients more control over their own health.

As we delve deeper into the complexities of cognitive computing and artificial intelligence, we are better able to comprehend the capacity of these technologies to enhance the capacities of humans and revolutionize entire sectors. We create the framework for a deeper investigation of how these technologies are transforming smart cities and revolutionizing healthcare in the future sections by promoting an awareness of cognitive systems and the diverse applications they have.

13.3 Increasing the Capabilities of Smart Cities Using Cognitive Computing

When cognitive computing is combined with the idea of smart cities, it opens up possibilities that have never been explored before for the creation of urban settings that are smarter, more adaptable, and more environmentally friendly. Cities have the ability to expand their capacities and provide citizens with a higher quality of life by incorporating cognitive technology into the current fabric of the infrastructure that makes up a smart city.

13.3.1 Cognitive Data Analytics for Smarter Cities

Cognitive computing provides smart cities with the ability to analyze massive volumes of data created by interconnected devices, which is more frequently referred to as the Internet of Things (IoT). Cities have the potential to glean useful insights into a variety of aspects of urban life, such as patterns of transportation, energy use, and environmental conditions, through the utilization of cognitive data analysis. The first step in the process is the methodical collection and integration of data from a wide variety of sources, including as Internet of Things devices, sensors, social media platforms, and public records. This will lay the groundwork for cognitive data analysis. Advanced AI-driven techniques, such as machine learning, deep learning, and predictive analytics, analyze the data to uncover hidden patterns and trends, providing a holistic understanding of infrastructure, transportation, energy consumption, environmental conditions, public services, and citizen behavior. These techniques include machine learning, deep learning, and predictive analytics. These invaluable insights equip city authorities and policymakers with the ability to make informed decisions and establish data-driven strategies, which ultimately leads to the development of innovative smart city apps. Cognitive data analysis improves the effectiveness of urban systems and services across the board, from traffic management to energy efficiency, thus contributing to the city's long-term viability and the citizens' overall quality of life. Cities have the potential to make strides toward being more resilient, sustainable, and citizen-centric if they adopt cognitive data analysis. This will pave the way for a future in which cities are smarter and more connected. These insights make it easier to make well-informed decisions and give municipal authorities the ability to optimize resource allocation in an effective manner.

13.3.2   Predictive Maintenance and Proactive Services

The possibility for predictive maintenance is one of the significant benefits that cognitive computing brings to smart cities. Cognitive systems are able to foresee the possibility of failures in essential infrastructure, such as transportation networks and utility systems, which enables preventive maintenance to be performed and helps to minimize expensive interruptions [5]. Cognitive computing may also improve service delivery in sectors such as waste management, public safety, and healthcare by evaluating historical data and real-time inputs. This makes cities more responsive to the demands of their citizens.

13.3.3   Personalized Urban Services

Thanks to cognitive computing, smart cities are able to provide individualized services that are catered to the tastes and requirements of each resident. Cognitive systems are able to provide users with individualized recommendations for transit routes, entertainment options, and local services because they are able to comprehend the behavior and preferences of users. This paradigm change is made possible by advanced technologies such as data analytics, artificial intelligence, and gadgets connected to the Internet of Things (IoT), all of which collect and handle huge volumes of data originating from a variety of different sources. This paradigm change is made possible by advanced technologies such as data analytics, artificial intelligence, and gadgets connected to the Internet of Things (IoT), all of which collect and handle huge volumes of data originating from a variety of different sources. Through the examination of data trends and individual preferences, cities are able to provide services that are both individualized and efficient, thereby elevating the standard of urban living. Not only can personalized urban services boost citizen satisfaction and involvement, but they also promote inclusivity by meeting the varied needs of communities and giving underserved groups more say in decision-making. Nonetheless, it is very necessary to address concerns regarding privacy and data security in addition to ethical considerations in order to guarantee responsible deployment.

Embracing customized urban services is a crucial step toward establishing smart, sustainable, and citizen-centric communities that prioritize the well-being and quality of life of its people. These cities aim to reduce environmental impact and improve residents' overall quality of life. Through the examination of data trends and individual preferences, cities are able to provide services that are both individualized and efficient, thereby elevating the standard of urban living. Not only can personalized urban services boost citizen satisfaction and involvement, but they also promote inclusivity by meeting the varied needs of communities and giving underserved groups more say in decision-making. Nonetheless, it is very necessary to address concerns regarding privacy and data security in addition to ethical considerations in order to guarantee responsible deployment. Embracing customized urban services is a crucial step toward establishing smart, sustainable, and citizen-centric communities that prioritize the well-being and quality of life of its people. These cities aim to reduce environmental impact and improve residents' overall quality of life. This individualized approach improves the whole urban experience for inhabitants as well as visitors, building a better feeling of community and involvement in the process.

13.3.4   Cognitive Computing and the Role It Plays in Obtaining Energy Optimization

Cognitive computing is an essential component in the process of obtaining energy optimization, which is a primary emphasis of smart city initiatives. Cognitive systems have the ability to improve energy usage in buildings and public spaces by studying patterns of energy consumption and data regarding the weather. This can result in lower carbon footprints and increased energy efficiency. Distributing energy throughout a city in a more consistent and dependable manner is another benefit that can be realized from the use of smart grids and cognitive energy management.

13.3.5   Data-Driven Decisions for City Development and Governance

Cognitive computing gives urban planners and policymakers the ability to make data-driven decisions on the development of cities and their governance. Cities have the ability to devise urban planning techniques that are more effective, to respond to emergencies in a more efficient manner, and to enact policies that address the changing demands of their populations by utilizing cognitive insights. The combination of cognitive computing with urban governance offers the framework for the development of smart cities [6] that are proactive, adaptable, and sensitive to the shifting dynamics of urban life.

The integration of cognitive computing technologies promises to significantly transform urban living as cities continue to evolve into smart and interconnected ecosystems. Cognitive computing enables cities to make strides toward a future of intelligent, sustainable, and inclusive urban settings through improving data analysis, enabling predictive maintenance, personalizing urban services, optimizing energy usage, and empowering urban government. The actualization of cognitively powered smart cities is a testament to the revolutionary capabilities of artificial intelligence in terms of transforming our societies for the better.

13.4   Cognitive Solutions Revolutionizing the Healthcare Industry

Cognitive solutions that are powered by AI technologies are undergoing a revolution in the medical field, where they are enhancing patient outcomes and medical practices while also undergoing a transformation in the way healthcare is delivered. The combination of cognitive computing and healthcare systems opens up new doors for data-driven diagnoses and individualized treatment of individual patients. Precision medicine can also benefit from these developments.

13.4.1 Artificial Intelligence-Driven Diagnostics and the Detection of Disease

Cognitive computing equips medical personnel with AI-driven diagnostics, which enables faster and more precise identification of disease. Cognitive systems are able to uncover patterns and indicators of diseases that may be difficult for human practitioners to recognize as a result of the analysis of extensive patient data, medical imaging, and genetic information [7]. It is possible that better treatment outcomes and increased chances of recovery for patients will result from early diagnosis that is helped by AI. Early disease diagnosis is improved by artificial intelligence's

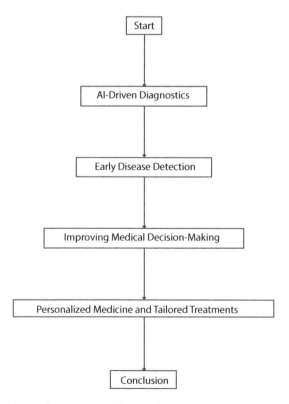

Figure 13.2 AI-driven diagnostics and disease detection.

unprecedented capacity to recognize subtle patterns and anomalies, and it has several procedures that is represented in Figure 13.2, which enables prompt interventions and perhaps reduces the expense of healthcare associated with late-stage therapies.

AI helps healthcare providers make informed and evidence-based treatment decisions, hence minimizing the likelihood of misdiagnosis and wasteful operations. This is accomplished through improving medical decision-making through data processing and generating insights from vast datasets. The implementation of AI in diagnostics paves the way for personalized medicine, which is a field of medicine in which treatment plans are individualized to specific patients based on the patients' individual features, such as their genetic profiles and lifestyle circumstances. Ethical issues, such as patient privacy, data security, and prejudice in algorithms, need to be deliberately addressed as artificial intelligence technology continues to improve in order to ensure responsible and equitable incorporation of the technology. Acceptance of AI-driven diagnostics and illness detection has the potential to usher in a new era of precision medicine in the healthcare industry, which would result in better patient outcomes and a shift in the way that medical professionals approach the delivery of healthcare.

13.4.2   Individualized and Tailored Treatment Programs

The capability of cognitive computing in healthcare to provide individualized treatment plans for patients is one of the most significant benefits offered by this type of technology. Cognitive systems are able to make individualized therapy recommendations for patients by taking into account patient features, medical history, genetic data, and treatment responses. These factors allow cognitive systems to offer treatment options that maximize effectiveness while minimizing adverse effects. This individualized approach not only improves patient satisfaction but also helps patients adhere more closely to their treatment plans.

13.4.3   Real-Time Monitoring of Patients and Predictive Analytical Tools

Real-time patient monitoring is made possible by cognitive computing; this involves continuously analyzing patient data in order to identify irregularities and shifts in the patient's overall health. This kind of monitoring makes early intervention and proactive care much easier to implement, which, in turn, helps cut down on hospital readmissions and improves patient outcomes. In addition, predictive analytics, which are powered by cognitive solutions, are able to forecast the occurrence of probable health concerns. This provides medical personnel with the ability to take preventative action and enhance the quality of patient care overall.

13.4.3.1   Cognitively Assisted Robotic Surgery

The application of cognitive computing in surgical settings is driving breakthroughs in robotic surgery. Surgeons are able to conduct complex surgeries with greater precision and dexterity when they integrate artificial intelligence algorithms with robotic devices. The cognitively aided robotic surgery cuts down on surgical errors, shortens the amount of time needed for patients to recuperate, and enhances patient safety, making it a significant advancement in the field of modern healthcare.

13.4.4   Patient Empowerment with Health AI

The use of cognitive computing not only alters the outlook of healthcare providers, but also provides patients with the ability to take responsibility for their own health. Patients are able to gain tailored insights into their health with AI-powered health applications and wearable devices, which encourages proactive health management. The patient population can become more involved and informed because of their ability to obtain real-time data on their vitals, receive advice regarding their health, and monitor their progress toward achieving their health goals [8].

A new era of patient-centered, data-driven, and precision-oriented medical care is about to begin, and it will be ushered in by the continued development of cognitive solutions and their integration with healthcare systems. Healthcare stakeholders can collectively provide individuals with higher quality care and improve the overall quality of their experience with healthcare by utilizing AI-driven diagnostics, personalized treatment plans, real-time patient monitoring, cognitive-assisted robotic surgery, and patient empowerment through health AI [9].

13.5   Application of Cognitive Computing to Smart Healthcare in Seoul, South Korea (Case Study)

Seoul, the busy metropolis of South Korea, is at the vanguard of the transformation of smart city healthcare, effortlessly integrating cognitive computing to alter its medical landscape. This places Seoul at the forefront of the smart city healthcare transition. The healthcare ecosystem in Seoul provides individualized and effective medical care while also improving urban health management. This is made possible via the utilization of technologies that are powered by artificial intelligence (AI) and insights that are driven by data. In order to discover illness patterns and risk factors, cognitive data analytics examines vast medical records, wearable devices, and health surveys. This paves the way for early identification and intervention that is specifically tailored to the problem. The city's AI-driven telemedicine services make it possible to conduct medical consultations remotely,

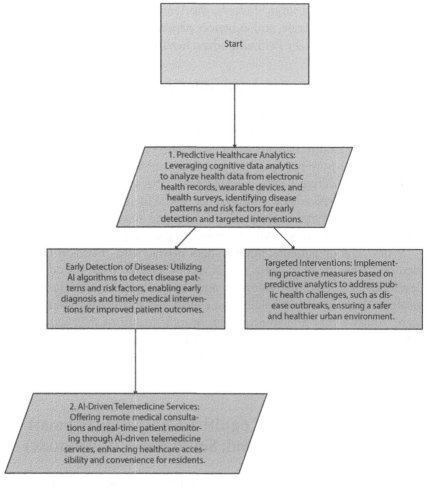

Figure 13.3 Seoul smart city healthcare: cognitive computing transformation.

thereby increasing both the availability and the convenience of healthcare. In addition to this, precision medicine customizes treatment plans to the unique characteristics of each patient, which results in improved therapeutic outcomes. By using a preventative rather than a reactive approach to the management of public health, Seoul is able to anticipate the spread of sickness and react appropriately and promptly. In order to guarantee a fair distribution of medical services, the allocation of resources is optimized using the insights gained from data analysis [10]. The experience that Seoul has had is illustrative of the revolutionary potential of cognitive computing. It has also established a wonderful precedent for other smart cities to follow in their quest of innovative and patient-centered healthcare solutions. The whole workflow can be seen in Figure 13.3.

13.6   Conclusion and Future Work

The incorporation of cognitive computing into the healthcare infrastructure of smart cities has shown to be a revolutionary force, bringing about a sea of change in the way medical procedures are carried out and improving the quality of life in urban areas. Cities have seen significant progress in patient care, precision medicine, and proactive public health management because of the utilization of technologies that are powered by artificial intelligence (AI) and insights that are driven by data. Residents living in a wide variety of areas have benefited from the increased accessibility and convenience of healthcare that has resulted from the introduction of AI-driven telemedicine services. The potential of cognitive computing in healthcare is still open to further investigation and innovation, even while the development of smart cities continues unabated. In the future, efforts may concentrate on developing AI algorithms to improve the accuracy of disease prediction and early detection, which would make it possible to take even more effective preventative measures. The expansion of data collecting and real-time patient monitoring is made possible by advancements in wearable technology and Internet of Things devices. This allows for continuous health tracking, which ultimately leads to better health outcomes. In addition, the combination of cognitive computing with new technologies such as blockchain has the potential to improve data security and privacy, which, in turn, could instill greater confidence in telemedicine services. In addition, the partnership that exists between cities and healthcare institutions can encourage the sharing of best practices and data, which, in turn, can contribute to collective breakthroughs in urban healthcare.

References

1. Wan, S., Gu, Z., Ni, Q., Cognitive computing and wireless communications on the edge for healthcare service robots. *Comput. Commun.*, 149, 99–106, 2020 Jan 1.

2. Srivani, M., Murugappan, A., Mala, T., Cognitive computing technological trends and future research directions in healthcare–A systematic literature review. *Artif. Intell. Med.*, 23, 102513, 2023 Feb.

3. Chen, M., Herrera, F., Hwang, K., Cognitive computing: Architecture, technologies and intelligent applications. *IEEE Access*, 6, 19774–83, 2018 Jan 15.

4. Coccoli, M., De Francesco, V., Fusco, A., Maresca, P., A cloud-based cognitive computing solution with interoperable applications to counteract illegal dumping in smart cities. *Multimedia Tools Appl.*, 1, 1–9, 2022 Jan.

5. Ewani, J., Bitala, W., Lotisa, V., Browndi, I., Smart city and future of urban planning based on predictive analysis by adoption of information technology. *J. Basis Appl. Sci. Manage. Syst.*, 4, 2022, 425–31, 2022 Jun.

6. Kontokosta, C.E. and Hong, B., Bias in smart city governance: How sociospatial disparities in 311 complaint behavior impact the fairness of data-driven decisions. *Sustain. Cities Soc.*, 64, 102503, 2021 Jan 1.

7. Bouchareb, Y., Khaniabadi, P.M., Al Kindi, F., Al Dhuhli, H., Shiri, I., Zaidi, H., Rahmim, A., Artificial intelligence-driven assessment of radiological images for COVID-19. *Comput. Biol. Med.*, 136, 104665, 2021 Sep 1.

8. Priya, R.L., Vaidya, A., Thorat, M., Motwani, V., Shinde, C., SAARTHI: Real-time monitoring of patients by wearable device, in: *2020 Advanced Computing and Communication Technologies for High Performance Applications (ACCTHPA)*, pp. 194–199, IEEE, 2020 Jul 2.

9. Lu, Z., Li, M., Annamalai, A., Yang, C., Recent advances in robot-assisted echography: Combining perception, control and cognition. *Cogn. Comput. Syst.*, 2, 3, 85–92, 2020 Sep.

10. South Korea prioritises AI-based imaging solutions to meet growing demand in hospitals, says GlobalData, ETHealthworld.com, Noida, India, 2022, November 24, https://health.economictimes.indiatimes.com/news/medical-devices/south-korea-prioritises-ai-based-imaging-solutions-to-meet-growing-demand-in- hospitals-says-globaldata/95734047.

14

Automating ESG Score Rating with Reinforcement Learning for Responsible Investment

Mohan Teja G.[1], Logesh Ravi[2], Malathi Devarajan[3] and Subramaniyaswamy V.[4]*

[1]School of Electronics Engineering, Vellore Institute of Technology, Chennai, India
[2]Centre for Advanced Data Science, Vellore Institute of Technology, Chennai, India
[3]School of Computer Science and Engineering, Vellore Institute of Technology, Chennai, India
[4]School of Computing, SASTRA Deemed University, Thanjavur, India

Abstract

This research presents a pioneering approach to calculating a company's ESG score rating, encompassing Environment, Social, and Governance criteria—a key method for assessing ESG performance. To achieve this, we integrate reinforcement learning, which enables an autonomous agent to make data-driven decisions and determine ESG scores based on multiple factors, including environmental impact, employee treatment, and governance standards. By employing a rigorous, transparent, and unbiased reinforcement learning procedure, we ensure fairness, prevent sustainability performance, and foster informed decision-making. The primary objective of this study is to foster a sustainable and responsible business environment by promoting consistency and transparency in ESG data and criteria. Our ESG rating model remains closely aligned with relevant news and contextual factors, generating more reliable scores for investors seeking to align their investments with sustainable values. Emphasizing accountability and sustainability, our methodology encourages companies across sectors to adopt more sustainable practices in their environmental, social, and governance policies. The novel approach presented in this chapter carries significant implications for the sustainable investment industry, providing investors with an invaluable, reliable, and impartial tool to assess a company's ESG performance, free from human biases.

*Corresponding author: vsubramaniyaswamy@gmail.com

Elakkiya, R. and Subramaniyaswamy V. (eds.) Cognitive Analytics and Reinforcement Learning: Theories, Techniques and Applications, (253–282) © 2024 Scrivener Publishing LLC

Furthermore, we conduct comparisons with other rating providers to illustrate the strengths of our data-driven methodology.

Keywords: ESG, reinforcement learning, sustainability, ESG score rating, responsible investment, transparency

14.1   Introduction

Environmental, Social, and Governance, or ESG, is a framework for evaluating a company's performance in sustainability. ESG is a term that was initially established in 2005. ESG covers many topics, such as supply chain management, organizational health, resource management, safety regulations, and transparency, which have an impact on financial implications on investments for the companies. The UN has released its sustainable development goals (SDGs) and they were widely starting to get adopted, which gave increasing attention to ESG practices [1, 2]. As these practices are seen as fundamental pillars of sustainability, the policy makers and governments, academia, and the general public have all taken notice of this [3–5]. Yet, there are wide variations in how ESG practices are implemented by corporations, and the lack of transparency and consistency in ESG data and methodology makes it difficult for investors to evaluate the sustainability performance of different businesses [6]. Due to the difficulties investors have evaluating the sustainability performance of different organizations due to the lack of transparency and standardization in ESG data and methods [6], this paper explores the benefits of investing in companies that prioritize sustainability, as evidenced by their ESG scores. By this analysis, we offer to investors an NLP model to help them make decisions that support sustainable enterprises and contribute to global change, with transparency of how the open system functions to help investors make a trustworthy decision.

a) ESG Practices and Significance

To determine a company's potential long-term value and its influence on stakeholders, ESG practices are very crucial [7]. Given the growing importance of sustainability in the business sector, research in this field will remain crucial for years to come. Some of the examples of factors that affect the financial viability that are covered by ESG practices are resources and supply chain management, organizational health, safety regulations, and transparency. The three ESG pillars—Environment, Social, and

Governance—have very distinct criteria from each other on which they are assessed.

Companies are assessed on their environmental stewardship under the Environmental pillar, which involves looking at things like greenhouse gas emissions, pollution, resource use, and land use. Companies are also obligated to report on any beneficial sustainability effects that might result in long-term business advantages. The supply chain norms, labor practices, product safety, and employee development are the main areas of the Social pillar. It also addresses problems with socially excluded groups having access to a company's goods and services. The Governance pillar, however, addresses issues including CEO compensation, shareholder rights, and company behavior, such as anti-competitive behavior and corruption.

The effectiveness of a company's corporate governance as well as its environmental and social impact can be evaluated using ESG ratings. ESG strategies can be used by businesses to measure and take actions that are profitable for the company, its stakeholders [7], and the environment. To evaluate the impact of ESG scores, we ran a survey that had 23 respondents from the ages of 19 to 23 as distributed in Figure 14.1.

These ages from 19 to 22 are the young investors who would start to invest as their career grows, but most of them never knew anything about sustainable investing, about 78.3% of them never heard of the ESG rating system, and only 22.3% of them did, as illustrated in Figure 14.2. Furthermore, all of them were interested in investing into sustainable companies as all of them replied that they were interested in investing into sustainable companies.

b) ESG Investing
ESG investing is a management strategy of investments that takes environmental, social, and governance factors into account when selecting investments.

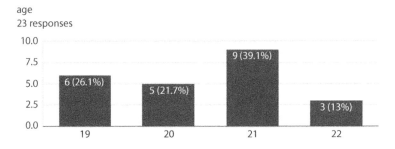

Figure 14.1 Distribution of ages of respondents in the survey.

Were you aware of the ESG scoring metric before this?
23 responses

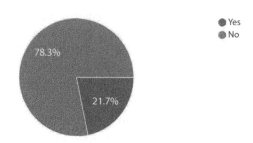

Figure 14.2 Percentage of the investors who are aware of ESG scoring metric.

It allows investors to support sustainable and socially responsible businesses while managing risk and potentially achieving high financial returns [10, 11]. While there are challenges to ESG investing, it has the potential to drive positive change and support a more sustainable and equitable future.

The objective of ESG investing is to identify companies that are sustainable, responsible socially, and governed well, while also avoiding companies that try to harm the environment, exploit their workers, or engage in unethical practices [17–19]. ESG investing has grown a lot in recent years as investors try to align their investments with their personal values and support companies that have a positive impact on society and the environment [8, 9].

One of the primary advantages of ESG investing is that it can promote positive change. Investors can help support sustainable practices, reduce environmental harm, and promote social justice by investing in companies that prioritize ESG factors [20]. An investor, for example, may prefer to invest in a renewable energy company that generates clean energy while reducing greenhouse gas emissions over a fossil fuel company that contributes to climate change. Similarly, the investor may prefer to invest in a company that promotes diversity, inclusion, and any other of his values in its workforce over one that engages in discriminatory practices.

ESG investing can also assist investors in risk management [8, 10]. Companies that prioritize ESG factors are more likely to be long-term sustainable and are less likely to face environmental or social crises that can harm their bottom line. For example, a company that invests in sustainable practices for farming is very less likely to suffer from crop failures due to climate change or soil depletion. Likewise, a company that promotes workplace safety and fair labor practices is less likely to face expensive lawsuits or reputational harm.

Despite its potential benefits, ESG investing is not without its difficulties. One of the most difficult challenges is defining and quantifying ESG factors. There is no standardized approach to ESG investing, and different investors may prioritize different factors based on their values and goals. One investor, for example, may prioritize environmental factors such as carbon emissions and resource use, whereas another investor may prioritize social factors such as labor practices and community engagement.

Another issue is a lack of transparency and standardization among businesses. There is no standardized system for reporting or verifying ESG data, and not all companies disclose their ESG practices and performance. This can make evaluating companies and making informed investment decisions difficult for investors.

Moreover, there is this widespread belief that ESG investing will always come at the expense of companies losing money, but this is not true, because numerous studies have found that companies that prioritize ESG factors perform really well financially in the long run; this is because these companies are often more innovative, resilient, and adaptable than their competitors, and are better positioned to navigate complex environmental and social challenges.

The demand for ESG investing is increasing rapidly in spite of these obstacles. The USSIF foundation made a report that states that the assets under ESG-mandated management has experienced a growth of 42%, growing from $12 trillion to $17.1 trillion from early 2018 to early 2020 [15]. This trend is most likely to continue as more and more investors are investing in sustainable businesses that align with their values. That has a positive impact on our environment and society. The consulting firm Deloitte predicts that ESG (environmental, social, and governance)-mandated assets will comprise 50% of all assets managed by professionals globally by the year 2024 [10].

c) Advantages of Sustainable Investing

Investing in companies that value and manage sustainability well, as evidenced by their Environmental, Social, and Governance (ESG) scores, has several benefits. First and foremost, it advocates for environmental preservation and positive social consequences. Second, when compared to investing in less sustainable businesses, investing in sustainable businesses has been shown to have lower financial risk and higher returns [7, 10]. Our survey also tried to find out the major reason for investing into sustainable companies, the results of which are depicted in Figure 14.3.

what is your main interest in investing in sustainable companies
23 responses

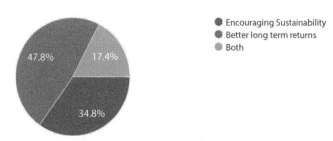

- Encouraging Sustainability
- Better long term returns
- Both

Figure 14.3 Major reason for investing into sustainable companies.

In addition, businesses that are sustainable benefit from cost savings and increased productivity of employees [12], either of which can help the business succeed in the long run. Sustainable businesses can save money by implementing energy-efficient practices, reducing waste, and using environmentally friendly materials [12]. Furthermore, businesses that are sustainable can attract and keep quality employees who share their values and are more enthusiastic about their work [16]. In the long run, these factors contribute to the business's stability and growth.

Moreover, funding sustainable businesses demonstrates a commitment to ethical standards and civic duty. It is becoming increasingly clear that companies with high ESG ratings are better able to adapt to changing market conditions and deal with economic upheaval. As a result, investing in sustainable businesses can provide long-term stability and growth.

To standardize the ESG reporting and to increase transparency in ESG data collection and analysis, efforts are being made at a global scale; several organizations and initiatives, including the Global Reporting Initiative (GRI) [17], the Task Force on Climate-related Financial Disclosures (TCFD) [18], the Sustainability Accounting Standards Board (SASB) [19], and the United Nations-backed Principles for Responsible Investment (PRI), are now working towards this goal.

These organizations have created frameworks and standards for ESG reporting that companies can use when disclosing their ESG practices and performance. The GRI standards [17], for example, offer a comprehensive set of guidelines for sustainability reporting, covering topics such as governance, ethics, human rights, labor practices, the environment, and social performance. The SASB standards offer industry-specific guidance on sustainability issues that are financially important to businesses [18]. The TCFD framework assists companies in disclosing climate-related financial risks and opportunities [19], whereas the PRI framework facilitates the incorporation of ESG factors into investment decision-making [21–23].

Aside from these frameworks, ESG rating agencies assess ESG performance of companies and provide ratings and rankings based on their findings. MSCI ESG Research, Sustainalytics, and Institutional Shareholder Services (ISS) ESG are examples of such organizations. These organizations assess companies' ESG performance using their own methodologies and criteria, and their ratings are frequently used by investors to inform their investment decisions.

Despite these efforts, ESG reporting and analysis still has room for improvement. The lack of standardization and consistency in ESG reporting is one issue. Companies may use various reporting frameworks or omit critical ESG information, making it difficult for investors to compare and assess their ESG performance. Additionally, ESG rating agencies may use different criteria and methodologies, leading to inconsistencies in their ratings.

Another issue is the possibility of green washing, which refers to the practice of companies making deals with such providers in order to make their company appear more appealing by having alleged deals with such companies, as evidenced by a lack of consistency and the reason for higher scores in some but low scores in others [24, 25].

In spite of these challenges, ESG investing is gaining traction and is expected to grow in popularity. Companies will face increasing pressure to improve their ESG practices and disclosure as more investors prioritize ESG factors in their investment decisions. As companies work to align their business strategies with sustainability goals, this can have a positive impact on the environment, society, and corporate governance.

While ESG scores cannot predict a company's stock market performance by themselves, they can be an important factor for investors to make decisions considering these factors. To test this model, we chose a few companies from the S&P 500 index, a stock market index composed of 500 of the largest publicly traded companies in the United States. The S&P 500 index includes companies from 11 different industry sectors. We calculated their ESG scores and presented them.

14.2   Comparative Study

The ESG rating agencies assess ESG performance of companies and provide ratings and rankings based on their findings. MSCI ESG Research, Sustainalytics, and Institutional Shareholder Services (ISS) ESG are examples of such organizations [26–30]. These organizations assess companies' ESG performance using their own methodologies and criteria, and their ratings are frequently used by investors to inform their investment decisions. In this section, the comparative study is made and presented.

a) Refinitiv ESG

Refinitiv, using publicly available data, provides ESG scores that assess a company's ESG performance across 10 important categories such as emissions, green product innovation, and human and shareholders' rights. The ESG ratings integrate industry materiality and business size biases and consider a transparent, data-driven evaluation of a company's relative ESG performance and capability. The ESG score by Refinitiv is calculated based on many company-level ESG indicators and many of the most comparable and relevant measures are used for overall company evaluation and grading.

Refinitiv's ESG measures consider factors such as inter-comparability, impact, availability of data, and relevance of the industry that differs across groups of industries. These measures are classified into 10 categories, which form the basis for the ESG scores. The ESG score is calculated using publicly available information and represents a company's ESG performance, commitment, and effectiveness. The scores from each category are combined to form the environmental, social, and corporate governance pillar scores, with varying category weights for the environmental and social categories depending on the industry. The governance category has the same weight across all industries.

b) Sustainalytics

The rating offers clear insights into ESG risk by assessing the organization's ESG that is not managed. This is evaluated using a distinct set of conditions, so it only considers issues that could significantly affect the economic value of the company. More than 14,000 companies are given risk levels in the ranking, ranging from minimal to severe. The three key factors that make up the ranking are corporate governance, environmental & social performance, and peculiar issues (black swans). They start with exposure, which is determined for each relevant ESG concern at the sub-industry level. The management component, which evaluates how effectively the organization reduces its exposure, is then looked at. Some businesses can believe that a portion of their risk is unmanageable. For instance, it is possible that an oil company will not be able to totally eliminate all of its risks related to carbon emissions; as a result, this is considered in the calculation. A company's policies, plans, practices, and quantitative performance measurements for the portion of risk that is manageable serve as indicators of its success. Because they show that the company's strategies and procedures have not been completely effective and could create risk, controversies lower the management score for the organization. The overall ESG Rating of a corporation is calculated by adding the amount of risk that is not managed for all relevant issues.

c) MSCI

MSCI ESG Ratings is a tool that assesses the long-term financial impact of ESG risks faced by companies. The goal of the ratings is to provide insight into a company's ability to withstand risks that may arise in the medium to long term. The ESG Ratings model answers four key questions about a company's ESG performance:

1) What are the primary ESG indicators to the company?
2) To which extent is the company vulnerable or positioned to benefit from these ESG risks and opportunities?
3) How effective is the company handling these ESG risks and opportunities?
4) How does the company's ESG performance measure up to that of its industry peers?

The final ESG Rating is calculated by taking the weighted average of individual Key Issue Scores, which are then normalized relative to other companies in the same industry. The ratings range from AAA (best) to CCC (worst) and take into account any adjustments made by the committee. The ESG Ratings model is industry-relative, meaning it takes into account each company's industry when assessing its ESG performance. Key Issue weights are set at the sub-industry level and are based on the industry's external impact and the time horizon associated with each ESG risk. These weights are reviewed and updated annually.

d) Bloomberg

Bloomberg's ESG Scores provide a clear link between each score and the company-reported data that drives it. The methodology behind the ES Scores, which can be found on BESG, includes details on the overall approach, industry-specific factors, quantitative methods, and all inputs to the model. Bloomberg uses a proprietary quantitative model that incorporates frameworks, research, and analysis to minimize noise, normalize data, correct for size bias, define peer groups, and summarize data into top-level scores. The analysis also separates sustainability performance from disclosure performance, which helps to penalize any gaps in sustainability performance and incentivize transparent disclosure.

e) S&P Global ESG Ratings

The S&P Global ESG Scores offer a comprehensive and distinctive method of assessing a company's ESG performance. In contrast to other ESG datasets that depend solely on publicly available information, the S&P Global ESG Scores are based on a variety of sources, including verified company disclosures, stakeholder analysis, media reports, and in-depth engagement through the S&P Global Corporate Sustainability Assessment (CSA). These scores provide a level of ESG insights that is unparalleled in the current market.

The first step in calculating ESG scores is the Corporate Sustainability Assessment (CSA), which involves talking to businesses and learning about their ESG policies. Companies invest hundreds of thousands of hours into each evaluation cycle, and S&P Global analysts review the disclosures for correctness and applicability, offer input, and discuss techniques and best practices. The CSA is the foundation of a privileged environment that propels corporate disclosures and develops sustainability standards over time.

When assessing companies, S&P Global uses the idea of "double materiality," taking into account both the financial and societal effects of ESG variables. They are aware of ESG variables that have a substantial financial impact on society or the environment as well as those that may have an influence on a company's value drivers, earnings potential, competitive posture, or long-term value for its shareholders. The ESG risks and opportunities for the company, including unreported and novel issues, are also taken into account. The industry-specific section of each questionnaire developed by S&P Global for the 61 sub-industries they cover uses industry-specific materiality matrices.

f) Moody's ESG

Moody's ESG Solutions is seeking feedback on proposed enhancements to its Environmental, Social, and Governance (ESG) Assessment methodology. The current methodology is based on a "double materiality" approach that considers both how ESG factors impact enterprise value and how a company's activities affect society and the environment. The proposed enhancements aim to improve the methodology by introducing new sub-categories for analysis, refining the double materiality approach, increasing the number of industry frameworks, enhancing the data structure, and adding an overall ESG grade. The new sub-categories include Physical Risks, Cyber & Technology Risks, and Responsible Tax, which will provide a more comprehensive and transparent analysis of a company's ESG performance. The refinements to the double materiality approach will balance shareholder and stakeholder interests, while the increased number of industry frameworks will ensure that ESG factors are appropriately weighted. The enhanced data structure will offer greater transparency into the scoring methodology and enable users to quickly understand a company's ESG performance.

14.3   Literature Survey

Overall, the survey suggests that ESG investing and rating is a rapidly growing market, with the potential for both financial and social benefits [32]. However, there are challenges in the current ESG rating products, such as bias and lack of standardization, which make it very difficult for potential investors to make informed decisions [33, 34]. The papers also indicate that companies with strong ESG performance tend to have higher firm value, and that disclosure of ESG activities is an important factor in understanding ESG performance and its relation to the firm value. Given these findings, the paper aims to address these challenges by developing a robust, unbiased method for determining a company's ESG score, providing a valuable tool for investors and other stakeholders to make informed decisions. Additionally, it is important that the project also keeps into account the evolution of ESG criteria and how these criteria are being used by ESG rating agencies and how these agencies are integrating sustainability principles into their corporate sustainability assessment process. This will help to ensure that the ESG scores generated by the project are accurate and relevant to the current market.

a) Investors' Valuation of Sustainability
A study examined whether investors take sustainability into account while making investment decisions; this study was published in the *Journal of Finance*. They discovered that funds with higher sustainability ratings saw an increase in fund flows, whereas funds with lower sustainability ratings saw a decrease in fund flows of over $12 billion [31]. The idea that investors do not care about sustainability information or penalize funds for keeping a portfolio of sustainable investments was rejected by the authors. A psychological research on categorization was conducted and the study found that investors primarily respond to simpler and more obvious global ratings rather than specific information on the percentile rank of sustainability within Morningstar categories.

The authors offered a number of justifications for the reactions to the disclosure of sustainability ratings, including societal limitations and non-financial factors, all of which have an impact on investors' choices. Contrary to the assumptions, more sustainable funds perform better and are more in line with the evidence. The authors came to the conclusion that a significant factor in investing decisions is the effect heuristic, which states that investors' feelings about an investment influence their expectations of future returns and perceived risk.

b) How Good Are the Raters

The expansion of investing sustainably and responsibly over the past 10 years is discussed in the paper, as well as the creation of ESG rating companies that analyze companies and measure their performance in sustainability with a variety of research approaches. The study seeks to address issues such as whether the ESG assessment criteria have changed over time, if the notion of ESG has become more consolidated in the previous 10 years, and whether the rating agencies are helping to promote more sustainable development [32].

The research examines how ESG rating agencies interact with society at large in addition to their impact on business. The study focuses on how ESG rating agencies incorporate sustainability into their evaluation frameworks and if they do it in a way that adheres to the key sustainability principles. These concepts include the intergenerational view, the stakeholder approach, and life-cycle thinking. They also contain the sustainability components of financial, economic, environmental, and social factors, as well as the balance between them.

The report conducts a comparative analysis of the financial market's most representative ESG rating and information provider agencies in order to conduct its examination. The research first examines how their evaluation standards have changed over the last 10 years, and then it looks at how ESG rating agencies have contributed to sustainable development.

The paper's major conclusions show that over the last 10 years, ESG rating agencies have improved their evaluation standards and incorporated sustainability principles into their evaluation procedures. The report does, however, also point out areas that still need work, notably in terms of stakeholder participation and life-cycle thinking. The research comes to the conclusion that, given the increasing importance placed on tracking corporate sustainability performance, the findings of this analysis are particularly relevant for understanding the behavior of these organizations and their evaluation processes [32].

c) How Much and Why Do ESG Ratings Differ Between Raters

Evaluating companies based on their environmental, social, and governance performance is becoming very popular among investors, regulators, and other stakeholders. ESG ratings, produced by specialized rating agencies, are used to assess a company's sustainability practices and identify potential risks and opportunities for investors. Anyhow, regardless of the demand for ESG ratings, there is a lack of standardization and consistency in these ratings.

To shed light on this issue, this study examines the divergence in ESG ratings among six prominent rating agencies: Kinder, Lydenberg, MSCI, Domini, S&P Global, Refinitiv, and Moody's ESG. By analyzing the agencies' methodologies using a common taxonomy, the study identifies the sources of divergence and quantifies their contribution to the overall variation in ratings.

The results show that measurement is the largest contributor to the divergence in ESG ratings, accounting for greater than half of the variation. Measurement refers to the way in which the agencies define and measure the indicators used to assess a company's ESG performance. The study finds that different agencies use different metrics and data sources to measure the same indicators, leading to divergent ratings.

Scope refers to the breadth of ESG issues considered by the agencies, which accounts for a little less than half of the divergence. Some agencies may focus on specific ESG issues or industries, while others take a more holistic approach, resulting in different ratings for the same company.

Weight, which refers to the relative importance assigned to different ESG issues, accounts for only a very small percentage of the divergence. However, the study finds that even small differences in weighting can result in significant differences in overall ratings.

In addition to these sources of divergence, the study identifies a rater effect, whereby a rater's overall perception of a company influences their rating of specific categories. For example, if a rater views a company as generally socially responsible, they may assign higher ratings to specific social indicators, regardless of the company's actual performance in those areas.

Overall, the study highlights the need for greater transparency and standardization in the data and methodologies used to generate ESG ratings. This is important not only for investors and other stakeholders who rely on these ratings, but also for companies who may be unfairly penalized or rewarded based on the subjective views of individual raters. By addressing the sources of divergence identified in this study, ESG rating agencies can improve the reliability and usefulness of their ratings, and contribute to a more sustainable and responsible business environment.

d) Does ESG Produce Value
The goal of this study is to look into the connection between a company's worth and its environmental, social, and governance actions and disclosure. The authors utilized a statistical model to investigate the relationship between these elements and firm value after gathering information on ESG ratings and whether the company provided ESG reports. According to the study, businesses with great ESG performance typically have higher firm values than those with a mediocre performance. The study also discovered that the mere disclosure of ESG operations has a tendency to lower firm value. The study did discover, however, that disclosure is essential for both reducing the impact of ESG deficiencies and enhancing the impact of positives. This study was conducted by Ali Fatemia, Martin Glaumb, and Stefanie Kaiser and published in ScienceDirect. They tried to model their model as Value of Firm = f(ESG Performance, ESG Disclosure, ESG Performance × ESG Disclosure).

14.4 Methods

The evaluation of a company's Environmental, Social, and Governance (ESG) performance is critical in today's world, as investors, customers, and employees demand greater transparency and accountability from businesses [13]. However, existing ESG rating systems have been criticized for their lack of accuracy and transparency, often based on subjective criteria that can be influenced by bias [14]. To address these issues, this paper has been developed that uses a data-driven and unbiased approach to evaluate a company's ESG performance. The methodology of this project involves using multiple data sources, including news articles from NewsAPI and publicly available metrics on the company's performance, such as carbon usage and employee feedback on diversity and inclusion, work–life balance, and more, scraped from Glassdoor. These sources are used to provide a comprehensive and unbiased view of the company's ESG performance. Figure 14.4 depicts the block diagram of the proposed methodology.

To ensure that the data that are collected are accurate and are free from bias as humanly possible, we used various natural language processing (NLP) techniques to process and clean the data for any outcomes that are not needed and weeded out. For example, SpaCy's classification model and named entity recognition are used to weed out any occasional ads that may be present in the data and to avoid the use of synonyms of the company's name, which may have another meaning. For example, Amazon is on fire here. Amazon is being referred to as the forest not the company, and ads

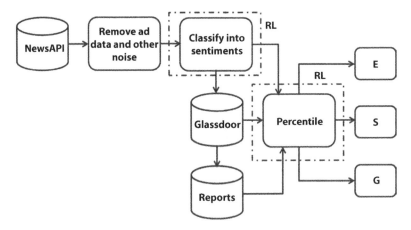

Figure 14.4 Block diagram of our methodology.

such as "Amazon great Indian sale is here" should be removed as they pose no importance to decide. These techniques help to ensure that the correct data are used in the analysis.

To differentiate between the two meanings of "Amazon," we used SpaCy's NER to identify whether "Amazon" is being used as an organization or as a location. The output tells us that SpaCy has identified "the Amazon" as a location (LOC) entity (this means that "Amazon" in this context refers to the Amazon rainforest) or "Amazon" as an organization (ORG) entity (this means that "Amazon" in this context refers to the company); this will help us classify them appropriately.

Furthermore, we used SpaCy's few shot classification task. Few-shot classification is a type of machine learning task that involves training a model to classify data into one of several categories based on a small number of training examples per category. This is in contrast to traditional machine learning classification tasks, which typically require a large amount of labeled data to train the model effectively. In a few-shot classification task, the model is trained on a few examples from each category, typically ranging from one to a few dozen examples per category. The goal is to develop a model that can generalize to new examples and accurately classify them into one of the known categories, even though it has seen very few examples from each category during training. Few-shot classification is particularly useful in situations where it is difficult or expensive to collect large amounts of labeled data. For example, in the field of computer vision, few-shot classification has been used to recognize novel objects with very few training examples, or to learn to recognize new classes of objects without having to retrain the entire model from scratch. We used our custom dataset of news articles for this task with hand-chosen ads and articles to train the model.

We chose to use SpaCy for these tasks because, for open-source natural language processing in python, SpaCy is designed to be efficient, fast, and easy to use, and to make NLP an excellent tool for processing large amounts of text data in our specific use case. One of the main benefits of SpaCy is its speed. It is optimized to be fast and can process large volumes of text data quickly, making it ideal for use in production environments. Additionally, it includes many pre-built models for common NLP tasks such as named entity recognition, part-of-speech tagging, and dependency parsing, which can save time and effort for developers and researchers.

SpaCy also offers a wide range of customization options, allowing users to train their models on custom datasets or to modify existing models to better fit their needs. We also tried to design it with a very user-friendly API structure in mind. This made it highly accessible to both experienced and novice users.

Employee feedback and ratings are scraped from Glassdoor using selenium and the Chrome web-driver. This provides valuable insights into the company's culture, diversity, and overall employee satisfaction. Glassdoor has a large and active community, with over 50 million unique monthly visitors, 150 million reviews, and 2.4 million employer clients. By using these metrics, the project can provide a more accurate and comprehensive view of the company's ESG performance.

The processed data are then analyzed using the zero-shot classification model; predicting the class of input data that the model has not seen during the training phase is the job of zero-shot classification. This method uses a pre-trained linguistic model and is a type of transfer learning. It is especially useful when there is a lack of readily accessible labeled data. We provide the model with a prompt and a sequence of natural language text that describes the intended task in order to achieve zero-shot classification. The model does not need to be given any examples of the intended task when using zero-shot classification. This differs from single or few-shot classification, which calls for the provision of one or a few examples of the intended task, such as BART. BART is a potent pre-trained language model that uses a left-to-right decoder and a bidirectional encoder in a sequence-to-sequence architecture. For problems including natural language creation, translation, and comprehension, it was first introduced as a de-noising auto-encoder. In the pre-training job, the original phrases' sequence is randomly changed, and a text-in-filling approach is used to substitute long stretches of text with a single mask token. The refinement of BART for text generation, comprehension, discussion, question-answering, and summarizing tasks has proven to be especially successful using this method. On the GLUE and SQuAD datasets, BART's performance is equivalent to RoBERTa's with comparable training materials, and it has produced new state-of-the-art results on a variety of abstractive tasks with gains of up to 6 ROUGE.

We have also tried various other methods on a test dataset with made-up news of 300 rows, and human-determined labels from that BART have performed the best with 84% accuracy, so we chose BART for our project, We used BART to categorize the texts into features of ESG, such as air pollution, water pollution, and others. This provides a more detailed and nuanced view of the company's performance in various ESG categories. To increase the transparency and credibility of the ESG scores provided, the project uses the Global Reporting Initiative (GRI) [17] framework to align the categorization of the collected data with ESG criteria. This framework provides a standardized way of reporting on ESG performance, making it easier for stakeholders to compare companies' performance across different industries and regions.

Our study aimed to identify factors that influence ESG data collection and review among companies. We gathered information from various sources and compared their ESG practices to identify any gaps or differences. We created a list of factors for each E, S, and G score and examined their relevance in news articles. Through comparative analysis, we determined the relevant features and used them in our classification model. The model evaluates each company's risk score by calculating how many companies are performing better or worse in each factor, with 1 being the riskiest and 0 indicating no risk. We then averaged the scores of all the remaining factors within each category (E, S, and G) to obtain a final risk rating of 0-1 for each company. The output of the model is a percentile score from 0 to 100 (0–1) as presented in Figure 14.5. Most of the users in the survey responded as they would like the score to be represented that way rather than grades because it would be easier for them compare and understand.

The metrics we used for the classification are as follows: customer dissatisfaction, loss, bad employee care, profit, board controversy, decreased pollution, and increased pollution. We classify the news into these categories using the BART model. The news articles were each of size approximately 500, which consisted of news articles listed by various sources of the company in the past month.

By using a data-driven and unbiased approach, this project aims to provide a more accurate and transparent evaluation of a company's

How do u expect the scores to be given as
23 responses

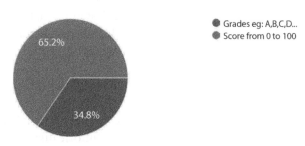

Figure 14.5 Percentile score.

ESG performance. This helps everyone including investors, customers, employees, and such to make a highly informed decision about which companies to support and invest their money into. The project's ultimate goal is to increase the transparency and credibility of ESG scores, making ESG evaluation a more trustworthy and reliable process.

NewsAPI is an essential tool used in this project to gather news data from over 80,000 worldwide sources. It is a simple, easy-to-use REST API that returns JSON search results for current and historic news articles. NewsAPI is trusted by over 500,000 developers, including those at Walmart, Accenture, Samsung, and Shopify. This demonstrates the reliability and accuracy of NewsAPI as a source of news data for the project. News API is available at a worldwide scale with hundreds of millions of articles in 14 languages from 55 countries, and is available for free during development and research.

Glassdoor is another critical tool used in this project to collect ratings and feedback on various metrics, such as diversity, CEO approval, and other ESG-related factors. Glassdoor is a thriving community for workplace conversations, with a simple mission to help people everywhere find jobs and companies they love. Since its start in 2007, Glassdoor has been a very proud champion of workplace transparency; they provide a platform for employees to share their valuable experiences and insights about their workplaces. With over 50 million unique monthly visitors, 150 million reviews, and 2.4 million employer clients, Glassdoor is a valuable source of data for the project.

We evaluated several companies, including Mercedes, Huawei, Microsoft, Google, Amazon, and Sony, by gathering 500 news articles for each for the past month. These companies were chosen because of the abundance of news articles available. The project then scraped the data from Glassdoor for each of these on aspects such as company rating, diversity rating, CEO approval, and recommend to a friend, and stored them in a CSV file. The project used various one-shot classification models on them, which gave similar results, of which the BART model was best to integrate and use.

The project cleaned and weeded out the news articles for potential ads and used NER to remove articles that were not related. After passing through the BART model, the project classified the news article into increased pollution, decreased pollution, board controversy, bad employee care, customer dissatisfaction, profit, and loss. Using these categories, the project calculated ESG scores of each company with features such as increased pollution, decreased pollution contributing to the Environmental pillar, bad employee care contributing to the Social pillar, and board controversy, customer dissatisfaction, profit, and loss contributing to the Governance pillar. These features can be updated whenever needed, and the project found these features to classify news articles well by observation.

Reinforcement learning is incorporated into this work to further enhance the process of calculating a company's ESG score rating. Reinforcement learning is a type of machine learning that involves an agent learning from interactions with an environment to achieve a specific goal. In this context, the goal would be to optimize the ESG scoring process and generate more accurate and reliable ratings. Reinforcement learning is applied to improve the ESG scoring methodology as described below:

- Define the Environment: In the context of ESG scoring, the environment can be represented as the dataset containing information about various companies and their corresponding ESG metrics.
- Agent and Actions: The agent in reinforcement learning is responsible for making decisions to achieve its goal. In this case, the agent can be designed to evaluate the ESG metrics of a company. The actions taken by the agent could be assigning weights to different ESG factors or selecting specific factors to include in the scoring process.
- Rewards and Goal: The goal of the reinforcement learning agent is to generate ESG scores that align well with real-world sustainability performance. To achieve this, the agent should receive rewards or penalties based on the accuracy and reliability of the ESG scores it produces. The rewards could be determined based on how well the ESG ratings align with external benchmarks or expert evaluations.

- Learning Process: The reinforcement learning agent learns from its interactions with the environment. It explores different strategies for calculating ESG scores, receives feedback in the form of rewards, and updates its decision-making process accordingly. Over time, the agent should converge to a more optimal method for generating ESG scores.
- Improvement Iterations: Reinforcement learning allows for continuous improvement through iterations. The agent can regularly update its ESG scoring approach based on new data and feedback, adapting to changes in the ESG landscape and investor preferences.
- Bias Mitigation: Since the reinforcement learning process is algorithmic and based on data rather than human decision-making, it can help reduce human biases in the ESG scoring process, ensuring a more impartial and transparent evaluation of companies' sustainability practices.

By incorporating reinforcement learning into the ESG scoring methodology, the approach has evolved and adapted over time, becoming more effective and reliable in evaluating a company's ESG performance. This can provide investors with an even more valuable tool for aligning their investments with sustainable values and promoting responsible business practices.

In the context of calculating a company's ESG score rating, reinforcement learning can be applied by defining the objective and agent: The objective of reinforcement learning in this context is to develop an algorithmic agent that can autonomously determine the ESG score for a company based on its environmental, social, and governance practices. The agent, in this case, could be a reinforcement learning model. The environment represents the dataset containing information about various companies and their ESG metrics. It includes data on a company's impact on the environment, treatment of employees, governance standards, and other relevant ESG factors. The actions in reinforcement learning refer to the decisions that the agent can take to achieve its objective. In this case, the actions would be the choices the agent makes to evaluate and assign scores to different ESG factors for a particular company.

The state represents the current situation or context in which the agent is making decisions. It could include the current ESG metrics of the company, historical data, and any other relevant contextual information. The observations are the inputs that the agent receives from the environment. For instance, it could be the specific ESG metrics and data available for the company being evaluated. In reinforcement learning, rewards are used to provide feedback to the agent based on its actions. In this context, the rewards could be determined by comparing the ESG scores generated by the agent with external benchmarks or expert evaluations. If the agent's ESG scores align well with real-world sustainability performance, it would receive positive rewards. Conversely, if the agent's scores deviate significantly from credible benchmarks, it would receive negative rewards (penalties). The reinforcement learning agent learns from its interactions with the environment. It explores different strategies for calculating ESG scores, takes actions based on the available data, and receives rewards or penalties as feedback. The agent's objective is to maximize the total reward it receives over time. Reinforcement learning allows for continuous improvement through iterations. The agent can update its ESG scoring approach based on new data and feedback received from previous evaluations. This iterative learning process enables the agent to converge to a more optimal method for generating ESG scores.

Reinforcement learning in the context of calculating a company's ESG score rating involves developing an algorithmic agent that learns to evaluate ESG metrics and assign scores to companies based on interactions with the dataset. The agent takes actions based on available information, receives rewards or penalties as feedback, and continuously improves its ESG scoring approach over time. By using reinforcement learning, the methodology can become more accurate, unbiased, and adaptable to changing ESG landscapes, ultimately providing investors with a valuable tool to make sustainable investment decisions.

14.5   Experimental Results

Environmental, social, and governance (ESG) ratings are becoming increasingly significant for investors in determining which companies to invest in. ESG ratings measure a company's performance in key areas related to sustainability, such as environmental impact, labor practices, and board diversity. As more investors seek to incorporate sustainability considerations into their investment strategies, the demand for reliable ESG ratings is growing.

The comparative study from various literature suggests that the market for ESG (environmental, social, and governance) investing and rating is experiencing rapid growth, driven by increasing demand from investors seeking to invest in companies that have the same value as them and to promote sustainable business practices. ESG ratings are designed to evaluate a company's performance in various areas, including environmental impact, labor practices, and ethical governance. However, current ESG rating products face several challenges that can hinder investors from making informed decisions.

One significant challenge is the potential for bias in the ESG rating process. Ratings can be influenced by subjective factors, such as the rating agency's interpretation of data or the weight given to different criteria. Furthermore, there is a lack of standardization in the ESG rating industry, with different rating agencies using different methodologies and criteria to evaluate companies.

Despite these challenges, research has shown that companies with a strong ESG performance tend to have a higher firm value, suggesting that ESG ratings can be a valuable tool for investors. Moreover, disclosure of ESG activities is an important factor in understanding the relationship between ESG performance and firm value. However, the lack of standardization and potential for bias in current ESG rating products can make it difficult for investors to assess a company's true ESG performance.

In this study, we analyzed the ESG ratings of six companies: Mercedes, Huawei, Microsoft, Google, Amazon, and Sony. We chose these six companies because they are major players in their respective industries and have a significant global presence. The data that we scraped from Glassdoor as metrics is presented in Table 14.1.

To collect the data we required for our analysis, we used a variety of sources, such as company reports, reviews of industry personnel, and, most importantly, the news articles. We focused on three key areas of ESG

Table 14.1 Glassdoor—sustainability report metrics.

Company	Recommend	CEO approval	Rating	Diversity	Self_ reports
Mercedes-Benz	85	89	4.2	4.1	0
Huawei	63	86	3.8	3.6	28
Microsoft	92	97	4.4	4.6	25
Google	89	89	4.5	4.4	0
Amazon	85	89	4.2	4.4	13
Sony	80	86	4.1	4	0

performance—environmental impact (E), social responsibility (S), and corporate governance (G)—and calculated an overall ESG rating for each company by averaging their scores in each area. This score here represents risk value from 0 to 1, i.e., the higher the value, the higher the risk. The results of our analysis are presented in Table 14.2.

Table 14.2 ESG scores of the companies.

Name	E	S	G	ESG
Amazon	0.53	0.43	0.29	0.41
Google	0.29	0.71	0.86	0.62
Huawei	0.51	1	1	0.83
Mercedes Benz	0.14	0.86	0.14	0.38
Microsoft	0.57	0.14	0.43	0.38
Sony	0.53	0.57	0.71	0.6

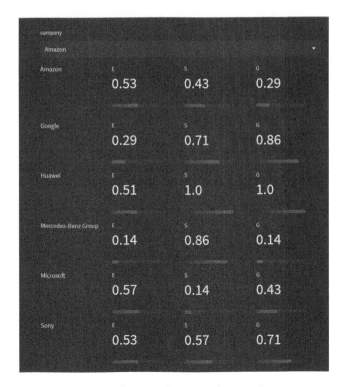

Figure 14.6 UI developed using the streamlit python framework.

We used a risk rating scale ranging from 0 to 1, where 1 represents the highest risk and 0 indicates the lowest; it is evident that Huawei poses a significant risk in terms of social and governance issues. This is due to recent events such as Germany's ban on Huawei's 5G smart phones and Huawei's decision to cancel a planned $1 billion research campus in Cambridge. These recent developments have contributed to Huawei's poor risk rating. However, as the data used to determine this rating only cover the last month, the model is highly sensitive to recent changes. While this sensitivity can accurately reflect recent trends, it may not consider past events adequately. To address this issue, collecting more data may help improve the accuracy of the risk rating.

We have conducted an in-depth analysis of the current status of ESG metrics provided by various companies and the relevant literature present on the subject. We have observed that although these scores have a significant impact and are highly regarded, there is widespread skepticism surrounding them. To improve their efficacy, the field requires more standardization, innovation, and transparency. We have also developed a UI

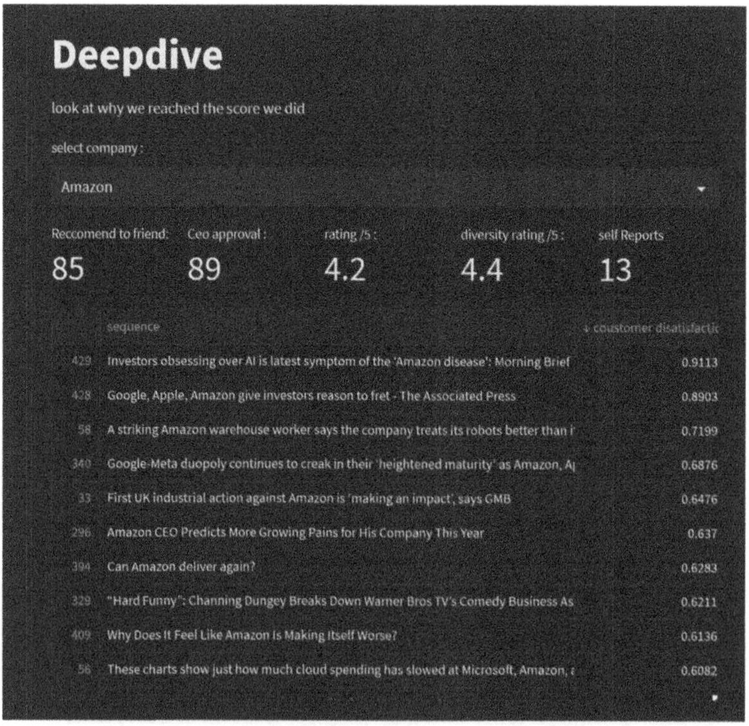

Figure 14.7 Deepdive image.

using the streamlit python framework, which could easily represent the same.

The ESG scores for the companies are presented in Figure 14.6 through the newly developed UI. It represents the risk ratings and ability to search through the company list, the searched one at the top and the remaining companies below. In Figure 14.7, we present all the news articles and data we used to get the score as a Deepdive image.

14.6   Discussion

The results of this study have several implications for investors and companies. First, our analysis suggests that our model provides a reliable estimate of a company's ESG performance. This is important because investors are increasingly using ESG ratings to make investment decisions, and they need accurate and reliable information to do so. Our model can serve as a useful benchmark for evaluating a company's ESG performance, and it can be adapted to include a wide range of companies in different industries and regions.

Second, our analysis highlights the importance of public perception in determining a company's ESG performance. Investors and companies alike should recognize the significance of maintaining a positive public image, as it can have a significant impact on their ESG ratings and ultimately their bottom line. Companies that prioritize sustainability and social responsibility initiatives are more likely to receive positive media coverage and attract socially conscious investors, which can result in increased investment and better financial performance.

Furthermore, our analysis shows that ESG performance varies widely across companies and industries. Some industries, such as technology and automotive, tend to have higher ESG ratings than others, such as oil and gas and mining. This suggests that investors should pay close attention to a company's industry when evaluating their ESG performance. Companies operating in industries with lower ESG ratings may need to take extra steps to improve their sustainability practices and enhance their public image.

Moreover, the literature suggests that ESG investing and rating is a growing market that can benefit both financially and socially. However, current ESG rating methods have problems such as bias and no set standards, which make it hard for investors to make good choices. Studies show that companies with good ESG performance usually have higher value, and sharing information about their ESG activities is important to understand the connection between ESG performance and value. This paper developed a fair and strong means to figure out a company's ESG score. This will help people make good choices about investing. It is also important to think about how ESG criteria are changing and how ESG rating groups include sustainability in their assessments. This will ensure that the ESG scores this paper makes are accurate and useful for today's market.

Furthermore, we can see the limitations of this mechanism of scoring for ESG metrics since Social and Governance scores are more publicly visible, and it is harder to get a grasp of the Environmental impact the company is having since this is a relative grading and since it is very rare for articles to talk about the Environmental impact of companies compared to those related to their Social and Governance standards; each article about them would get heavy weightage to prevent this, and the model could be further developed with newer data sources that will cover the Environmental impact of a company.

Finally, our study highlights the importance of transparency in ESG reporting. Investors need access to accurate and reliable information to make informed investment decisions, and companies that provide transparent ESG reporting are more likely to receive positive ratings from organizations like Sustainalytics. Companies that prioritize transparency and disclosure of ESG data are also more likely to identify areas for improvement and take steps to enhance their sustainability practices.

14.7   Conclusion

In conclusion, our analysis provides valuable insights into the significance of ESG ratings for investors and companies. Our model provides a reliable estimate of a company's ESG performance, which can be useful for investors who are looking to make informed decisions regarding sustainable investments. Our study also highlights the significance of public perception, industry, and transparency in ESG reporting. The public perception of a company's sustainability practices can have a significant impact on their ESG ratings and ultimately their financial performance. Companies that prioritize sustainability initiatives are more likely to receive positive media coverage and attract socially conscious investors, which can lead to increased investment and better financial performance. Moreover, our analysis indicates that ESG performance varies significantly across companies and industries. Companies operating in industries with lower ESG ratings may need to take extra steps to improve their sustainability practices and enhance their public image. This suggests that investors should pay close attention to a company's industry when evaluating their ESG performance.

Furthermore, our comparative study of the market suggests that there are challenges in the current ESG rating products, such as bias and lack of standardization, which can make it difficult for investors to make informed decisions. Companies with strong ESG performance tend to have higher firm value, and that disclosure of ESG activities is an important factor in understanding the relationship between ESG performance and firm value.

Finally, our study emphasizes the importance of transparency in ESG reporting. Investors do need access to a very accurate and reliable information source to make informed investment decisions. Companies that provide transparent ESG reporting are more likely to receive positive ratings from organizations like Sustainalytics. Moreover, companies that prioritize transparency and disclosure of ESG data are also more likely to identify areas for improvement and take steps to enhance their sustainability practices; although this model used data from news API for companies over the past month, incorporating additional information and conducting thorough data cleaning could significantly enhance the model's usefulness. Overall, our analysis suggests that ESG ratings are an essential component of sustainable investing, and companies that prioritize sustainability initiatives and transparent reporting will be better positioned to attract socially conscious investors and enhance their financial performance. As investors continue to prioritize sustainability considerations in their investment decisions, reliable and accurate ESG ratings will become increasingly important.

References

1. Alkaraan, F., Albitar, K., Hussainey, K., Venkatesh, V.G., Corporate transformation toward industry 4.0 and financial performance: The influence of environmental, social, and governance (ESG). *Technol. Forecasting Soc. Change*, *175*, 121423, 2022.
2. Duque-Grisales, E. and Aguilera-Caracuel, J., Environmental, social and governance (ESG) scores and financial performance of multilatinas: Moderating effects of geographic international diversification and financial slack. *J. Bus. Ethics*, *168*, 2, 315–334, 2021.
3. Mooneeapen, O., Abhayawansa, S., Mamode Khan, N., The influence of the country governance environment on corporate environmental, social and governance (ESG) performance. *Sustain. Account. Manage. Policy J.*, *13*, 4, 953–985, 2022.
4. Birindelli, G., Bonanno, G., Dell'Atti, S., Iannuzzi, A.P., Climate change commitment, credit risk and the country's environmental performance: Empirical

evidence from a sample of international banks. *Bus. Strategy Environ.*, *31*, 4, 1641–1655, 2022.

5. Wasiuzzaman, S., Ibrahim, S.A., Kawi, F., Environmental, social and governance (ESG) disclosure and firm performance: Does national culture matter? *Meditari Account. Res.*, 31, 5, 1239–1265, 2022.

6. Escrig-Olmedo, E., Fernández-Izquierdo, M. Á., Ferrero-Ferrero, I., Rivera-Lirio, J.M., Muñoz-Torres, M.J., Rating the raters: Evaluating how ESG rating agencies integrate sustainability principles. *Sustainability*, *11*, 3, 915, 2019.

7. Zumente, I. and Bistrova, J., ESG importance for long-term share- holder value creation: Literature vs. practice. *J. Open Innov.: Technol. Mark. Complex*, *7*, 2, 127, 2021.

8. Hartzmark, S.M. and Sussman, A.B., Do investors value sustainability? A natural experiment examining ranking and fund flows. *J. Finance*, *74*, 6, 2789–2837, 2019.

9. Vijayakumar, V., Vairavasundaram, S., Logesh, R., Sivapathi, A., Effective knowledge based recommender system for tailored multiple point of interest recommendation. *Int. J. Web Portals (IJWP)*, *11*, 1, 1–18, 2019.

10. Rahman, M., II and Lau, W.Y., How has ESG investing impacted investment portfolios? A case study of the Malaysian civil service pension fund. *Asian J. Econ. Modelling*, *11*, 1, 15–28, 2023.

11. Sankar, H., Subramaniyaswamy, V., Vijayakumar, V., Arun Kumar, S., Logesh, R., Umamakeswari, A.J.S.P., Intelligent sentiment analysis approach using edge computing-based deep learning technique. *Softw.: Pract. Exper.*, *50*, 5, 645–657, 2020.

12. Becker, M.G., Martin, F., Walter, A., The power of ESG transparency: The effect of the new SFDR sustainability labels on mutual funds and individual investors. *Finance Res. Lett.*, *47*, 102708, 2022.

13. Siew, R.Y., A review of corporate sustainability reporting tools (SRTs). *J. Environ. Manage.*, *164*, 180–195, 2015.

14. SIF, U., The US SIF foundation's biennial 'trends report' finds that sustainable investing assets reach $17.1 trillion. *US SIF: The Forum for Sustainable and Responsible Investment*, 2020, www. ussif. org.

15. Stubblefield Loucks, E., Martens, M.L., Cho, C.H., Engaging small- and medium-sized businesses in sustainability. *Sustain. Account. Manage. Policy J.*, *1*, 2, 178–200, 2010.

16. GRI reporting standards, a-short-introduction-to-the-gri-standards.pdf (globalreporting.org). https://www.globalreporting.org/media/wtaf14tw/a-short-introduction-to-the-gri-standards.pdf

17. Hales, J., Sustainability accounting standards board (SASB), in: *World Scientific Encyclopedia of Climate Change: Case Studies of Climate Risk, Action, and Opportunity Volume 3*, pp. 37–41, 2021.

18. Board, F.S., Task force on climate-related financial disclosures. Final Report: Recommendations of the Task Force on Climate-Related Financial Disclosures, 2017.

19. Chen, Z. and Xie, G., ESG disclosure and financial performance: Moderating role of ESG investors. *Int. Rev. Financ. Anal.*, *83*, 102291, 2022.
20. Public responsible investing framework PRI, PRI | Home (unpri.org). https://www.unpri.org/introductory-guides-to-responsible-investment/what-is-responsible-investment/4780.article
21. Yu, E.P.Y., Van Luu, B., Chen, C.H., Greenwashing in environmental, social and governance disclosures. *Res. Int. Bus. Finance*, *52*, 101192, 2020.
22. Selvan, N.S., Vairavasundaram, S., Ravi, L., Fuzzy ontology-based personalized recommendation for internet of medical things with linked open data. *J. Intell. Fuzzy Syst.*, *36*, 5, 4065–4075, 2019.
23. Logesh, R., Subramaniyaswamy, V., Malathi, D., Senthilselvan, N., Sasikumar, A., Saravanan, P., Manikandan, G., Dynamic particle swarm optimization for personalized recommender system based on electroencephalography feedback. *Biomed. Res.*, *28*, 13, 5646–5650, 2017, (0970-938X).
24. Refintiv ESG ratings score, https://www.refinitiv.com/en/sustainable-finance/esg-scores.
25. Subramaniyaswamy, V., Logesh, R., Vijayakumar, V., Indragandhi, V., Automated message filtering system in online social network. *Procedia Comput. Sci.*, *50*, 466–475, 2015.
26. Sustainanalytics ESG score, https://www.sustainalytics.com/.
27. MSCI ESG ratings, https://www.msci.com/our-solutions/esg-investing/esg-ratings?creative=638365660924&keyword=msci%20research%20esg&matchtype=p&network=g&device=c&gclid=Cj0KCQiA54KfBhCK-ARIsAJzSrdoKLnggso7nOsXLDICk2ZZ5-Q9_xiyuuIBnSm8RUru2KoqQB-wmFRRIaAuFeEALw_wcB&gclsrc=aw.ds.
28. Blooomberg ESG reporting, https://www.bloomberg.com/professional/dataset/global-environmental-social-governance-data/.
29. S&P Global ESG ratings, https://www.spglobal.com/ratings/en/products-benefits/products/esg-evaluation.
30. Moodys ESG rating, https://esg.moodys.io/.
31. Hartzmark, S.M. and Sussman, A.B., Do investors value sustainability? A natural experiment examining ranking and fund flows. *J. Finance*, *74*, 6, 2789–2837, 2019.
32. Escrig-Olmedo, E., Fernández-Izquierdo, M. Á., Ferrero-Ferrero, I., Rivera-Lirio, J.M., Muñoz-Torres, M.J., Rating the raters: Evaluating how ESG rating agencies integrate sustainability principles. *Sustainability*, *11*, 3, 915, 2019.
33. Berg, F., Koelbel, J.F., Rigobon, R., Aggregate confusion: The divergence of ESG ratings. *Rev. Financ.*, *26*, 6, 1315–1344, 2022.
34. Fatemi, A., Glaum, M., Kaiser, S., ESG performance and firm value: The moderating role of disclosure. *Glob. Finance J.*, *38*, 45–64, 2018.

15

Reinforcement Learning in Healthcare: Applications and Challenges

Tribhangin Dichpally, Yatish Wutla and Sheela Jayachandran*

Department of Computer Science and Engineering (SCOPE), VIT-AP University, Amravathi, Andhra Pradesh, India

Abstract

Reinforcement learning, also known as RL, is a decision-making approach that entails engaging with complicated settings with the goal of maximizing the long-term benefits while abiding by a predetermined policy and accepting helpful criticism for improvement. A computerized agent interacts with a novel environment, makes decisions, and eventually learns about its dynamics. It has proven advantages over other learning strategies when it comes to making medical decisions. It places a focus on long-term gains and is capable of handling difficult, lengthy sequential decision-making problems involving delayed, exhaustive, and sampled data. It has evolved into a potent strategy for creating effective healthcare solutions. In order to analyze the function of RL in healthcare, this chapter looks at earlier research, points out any flaws, and makes assumptions about potential future contributions.

Keywords: Reinforcement learning, Markov decision process, decision making, artificial intelligent, machine learning

15.1 Introduction

As a result of rapid advances in computer science and creation of sophisticated algorithms, the growing field of computing intelligence has achieved tremendous progress towards its ultimate goal. AI, which simulates intelligence on programmable devices, aims to duplicate the features and

---

Corresponding author: Sheela.j@vitap.ac.in

Elakkiya, R. and Subramaniyaswamy V. (eds.) Cognitive Analytics and Reinforcement Learning: Theories, Techniques and Applications, (283–316) © 2024 Scrivener Publishing LLC

capabilities of the human brain. Contrary to common belief, AI seeks to improve and complement human creativity, reasoning, and decision-making abilities rather than to completely replace humans with artificial workers. Machine learning (ML), one of the many subfields of AI, has become a potent technology that enables software programs to generate precise predictions without explicit programming. ML can be split into three major categories: unsupervised learning, supervised learning, and reinforcement learning. Each category focuses on a particular learning situation and uses unique methods to draw conclusions from the data.

A process similar to how individuals learn from their past experiences and pick up new information to improve their ability to carry out tasks in the real world is supervised learning. The foundation of supervised learning is the capacity to identify correlations and trends within input–output pairings [1]. With supervised learning, the model is explicitly given labeled samples, allowing it to learn relationships between input features and associated output labels. The model gains a broad understanding of the data distribution through this process, enabling it to forecast results for unknown inputs. The success of supervised learning algorithms is attributable to the accessibility of big labeled datasets, technological breakthroughs, and algorithmic innovations. These algorithms have outperformed humans in a number of fields, showing they have the potential to significantly increase automation, decision-making, and data-driven insights. Unsupervised learning is an ML paradigm that focuses on discovering underlying patterns and structures in unlabeled datasets. This approach is distinct from supervised learning, which uses labeled datasets to train machine learning models, and reinforcement learning, which focuses on maximizing a reward signal. While some may consider categorizing reinforcement learning as just an unsupervised learning type, this procedure involves an agent interacting with the external environment. The agent experiences consequences or rewards as an outcome of its interactions.

Reinforcement learning, as opposed to supervised and unsupervised learning [2], is a form of machine learning that relies on goal-directed learning. A machine is trained from a dataset that has previously been classified by an outside supervisor in supervised learning. Each example in the data collection gives a description of a specific scenario and a label that corresponds to the appropriate course of action to be used in the situation. The system uses these labels to determine which category the situation most closely fits. The purpose of this type of learning is to enable the machine learning system to appropriately generalize its findings, enabling it to perform as intended even in circumstances that are not included in the training set. Although it is believed that this type of learning is very significant, it is insufficient for learning by touch. When there are interaction issues, it is usually challenging to compile examples of the required behavior that are both precise and inclusive of all the situations where the agent must behave. As a result, the agent must possess the ability to learn from its own experiences in unfamiliar territory, where learning is perceived to be the most beneficial. This requires the development of advanced algorithms that enable the agent to learn and adapt in real time. The use of machine learning and other advanced computational techniques can facilitate the development of such algorithms and enable agents to learn from their own experiences, even in previously unexplored terrain.

Exploring the various methods that use Reinforcement Learning within the healthcare domain is the main objective of this chapter. In particular, the objective is to explore the possibility of employing this approach to assist doctors in making informed decisions by providing them with easier and faster access to relevant information while ignoring redundant or irrelevant information. In order to achieve this objective, researchers have employed treatment plans for a variety of disorders that can be optimized using reinforcement learning and deep reinforcement learning techniques. These methods enable medical professionals to provide their patients with the best possible care and make better treatment decisions. They include the delivery of anticoagulant medications [3], anticoagulant treatment regimens [4], sepsis treatment [5], pharmacological therapy for myeloma patients [6], and HIV (human immunodeficiency virus) medication regimens [4].

15.2   Structure of Reinforcement Learning

Through a cycle of collaboration between the agent and the environment, reinforcement learning aims to find the optimum strategy that maximizes the cumulative reward over time. The agent receives a current state observation from the environment, which starts the reinforcement learning loop. Using its present policy, the agent then decides on a course of action. The environment receives the action signal and replies with a new state and a signal of reward. Based on the changed state and the received incentive signal, the agent subsequently adjusts its policy. This approach is done until the agent has discovered the best policy. The Markov decision process (MDP) [7] is an equation-based structure for simulating decision-making scenarios whose results are influenced to some extent by arbitrary events. It may be used to represent reinforcement learning systems. MDPs presuppose that the present state of the system provides all the necessary data to make the optimal choice. A partly observable Markov decision process (POMDP) is a more complex problem that arises when the state of the system is not instantly visible in some circumstances [8].

a) Markov Decision Process (MDP)
This process is a tuple ⟨State, Action, Transition, Reward⟩ in which State is a limited number of states, Action is a limited number of actions; Transition is a function defined as:

Transition: State × Action × State → [0,1]
Reward is a function defined as:
Reward: State × Action × State → Reward [7].

The Markov Decision Process is a popular paradigm for modeling decision-making processes in a stochastic environment. It is only a reinterpretation of Markov chains, but it includes a decision-making mechanism that makes it possible to map the best course of action for each individual environment state [9]. According to the Markovian property, which is the basis for MDP, only the current situation is significant for predicting the future. This demonstrates that the probability distribution of the following state is solely determined by the present state and the action executed. As a result, MDP may be used to simulate systems that display this trait, such as many real-world systems involving uncertain decision-making.

b) Partially Observable Markov Decision Process (POMDP)
A partially observable Markov decision process (POMDP) is a tuple (State, Action, Trans, Observe, Reward, γ), where

- A state is a limited number of states,
- An action is a limited number of actions,
- Trans is a transition function that indicates the likelihood of transitioning from state x1 to state x1' when action a is taken: $Trans(x1, a, x'1) = P(x1' \mid x1, a)$.
- Observe is an observation function that specifies the likelihood of receiving observation O when action a is taken in state x and true state is x': $Observe(x, a, O) = P(O \mid x, a)$.
- The reward is a function that describes instantaneous reward $r(x, a)$ obtained when action a is taken and is carried out in state x.
- γ is the discount factor that defines how significant immediate benefits are in comparison to future ones [10].

The environment is described by a POMDP's Markov Decision Process. It consists of a number of states, a number of operations, and a transition function that illustrates how likely it is that the current state changes whenever an action is performed. The agent, however, is unable to access the environment's actual state, unlike in an MDP [11]. Instead, the agent receives observations that are related to the true state of the environment in a probabilistic way. In a POMDP, the agent's purpose is to choose a series of acts that maximizes a long-term reward function. To accomplish this, the agent needs to maintain a probability distribution that is a belief state across the potential states of the environment based on the agent's previous observations and actions. The agent's uncertainty regarding the true condition of the environment is represented by the belief state. Finding an optimal policy, which is a function that maps each belief state to an action, is required to solve a POMDP. POMDPs can be solved using different algorithms, such as value and policy iterations. POMDPs are employed in a wide range of applications, including robots, self-driving cars, and healthcare. They provide an effective framework for modeling complicated decision-making situations with uncertainty and partial observability. To clearly understand Reinforcement learning, we need to understand the terms agent, environment, state, reward, and action, represented in Figure 15.1.

c) Agent

Typically, the agent is portrayed as a decision-making body that receives observations from the environment, takes actions in response, and receives a reward signal indicating how well it is performing. Finding a strategy that maximizes the anticipated cumulative benefit over time is the agent's goal; it acts as a map connecting states with actions. In RL, there are various types of agents, each with their unique set of strengths and drawbacks. Among the most prevalent types of agents are:

Model-based agents: These agents maintain an environment model in order to forecast the results of certain actions. Model-based agents have

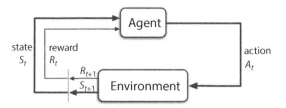

Figure 15.1 Reinforcement learning process.

the potential to be more efficient than other types of agents because they can prepare ahead of time and avoid activities that are likely to result in bad outcomes [12].

Model-free agents: These agents develop a policy by experience rather than maintaining a model of the environment. Model-free agents have the potential to be more adaptable than model-based agents since they can change to the environment without needing to build a new model.

d) Environment
For each RL challenge, we must create the appropriate environment. A collection of rules that outline the things an agent is allowed to perform, the states that the environment contains, and other factors make an environment and the applicable rewards and punishments. The term "environment" describes the physical place in which a piece of software or an agent can function or communicate. The environment's reward and following state are its outputs, and an agent's action and current state are its inputs. The environment (reward, resultant state) is responsible for everything that influences and governs how an agent acts and the results of those behaviors.

e) State
An agent's reachable states, locations, and positions inside a specific environment are referred to as states. In order to describe a state, one can use coordinates, numerals, or alphanumeric characters. Both a finite and an infinite number of states are possible. Non-transitioned out or process-terminating states are referred to as terminal states. An agent makes a visit when they go from one state to another, and an episode is when they make several trips.

f) Reward
When an agent completes an activity, the Reward feature rewards them with a numerical value for being in that state. Because rewards signal if a state is advantageous or not, agents are rewarded more when they move through advantageous states and less when they do so. A reward can give an agent either good or bad criticism. A reward's character or value can affect how people act and how policies are implemented. Little changes can have a big impact on how agents behave.

g) Action
Anything that a human or a robot can think of doing or is permitted to accomplish in a given environment is considered an action or a series of

actions. An agent can move left, right, up, down, or remain in the current position in a grid environment, for instance. An agent has left and right movement options in a single horizontal line environment. Generally speaking, an agent may select any action that is offered in a particular context.

15.3   Applications

15.3.1   Treatment of Sepsis with Deep Reinforcement

Sepsis is among the primary causes of mortality in intensive care facilities, and hospitals treat it with billions of dollars each year. Since every patient responds to medical interventions quite differently and because sepsis has no known cure, treating a septic patient is particularly difficult. Our strategy uses continuous state-space models with deep reinforcement learning to provide therapy suggestions for septic patients. The work that addressed the problem of a sepsis treatment strategy to identify treatment regimens for septic patients used modeling with discretized action and state spaces. Value-iteration techniques were employed in this study to identify the ideal policy. To simulate time-varying state spaces in reinforcement learning, processes are often utilized.

The authors of the article employed a subset of sepsis of the MIMIC-III dataset. The data for this cohort came from the Multiparameter Intelligent Monitoring in Intensive Care (MIMIC-III v1.4) database, which concentrates on individuals who meet the Sepsis-3 criteria (17,898 in total). Each patient's demographics, test results, vital signs, and intake/output events were recorded, along with other pertinent physiological characteristics. They decide to say that Vasopressors and IV fluid make up the action area. The four bins that make up the division of pharmacological dosages each include a different dosage of each substance. The action space was divided into quartiles for each of the two medications based on all non-zero dosages. Figure 15.2 shows how the various model learned the policies.

At each time step, each drug was then converted into an integer to represent its quartile bin, with bin 0 acting as a special case when no drugs were present. Double-Deep Q Networks, which have separate value and benefit flows, make up the core network. Competing Double-Deep Q, the inquiry employs networks to approximate the ideal action-value function, Q, and discretized action space/fully continuous state-space models to discover the optimum treatment options (s,a). The SOFA (sequential organ failure assessment) score, which evaluates organ failure and patient lactate levels

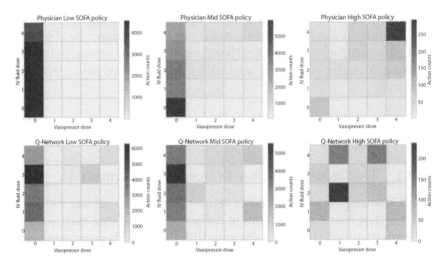

Figure 15.2 Policies learned by the various models.

as markers of a patient's general health, provides the basis for the clinical motivation of the reward function. Higher SOFA scores, spikes in SOFA scores, and lactate levels are all penalized by this reward mechanism. For evaluation, they use the U-Curve. They emphasize how the dosage disparity between the advised policy and the actual policy affects the mortality rate. Continuous state-space modeling has proven to be capable of revealing interpretable rules that help improve clinical practitioners' treatment recommendations. Figure 15.3 compares predicted doses and dosage recommended by experts. Future research should investigate various modeling techniques and look at the learned procedures patient by the patient (such as model-based reinforcement learning) [13].

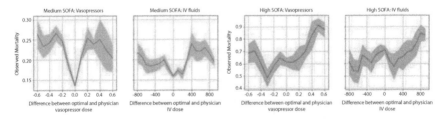

Figure 15.3 Comparison of the difference in doses between those recommended by the ideal policy versus those prescribed by doctors (x-axis) and the observed mortality (y-axis).

15.3.2   Chemotherapy and Clinical Trial Dosing Regimen Selection

In RL domains, reward functions are often derived from measures that are independent of the representation of states and actions chosen. While many existing domains in RL have well-defined outcomes, utilizing RL to solve issues in the absence of an easily accessible external scalar outcomes remains relatively unexplored (Sutton and Barto, 1998) [14]. To achieve the maximum reduction in patients' mean tumor diameter (MTD), they developed an RL framework to systematically explore dosage options. The model framework is displayed in Figure 15.4. By employing innovative state- and action-based reward functions with various trade-offs between actions and outcome proxies, they learned dose regimens in clinical trials involving vincristine (PCV), temozolomide (TMZ), and 1-(2-chloroethyl)-3-cyclohexyl-l-nitrosourea and procarbazine. This demonstrated the partitioning of the trial design space. To facilitate the application of reinforcement learning (RL) in domains where external rewards are limited and unstructured, they proposed a method that incorporates action-derived rewards.

In TMZ and PCV chemo- and radiotherapy (CRT) studies, learned dosing regimens significantly lower MTD for cohorts and individual patients while requiring lower drug concentrations and dosages than human-expert dosing regimens. Rewards based on patient outcomes are particularly common in learning issues in the medical field. For instance, current research employing RL in intensive care units solely rewards patient survival rates instead of the domain's state and action spaces.

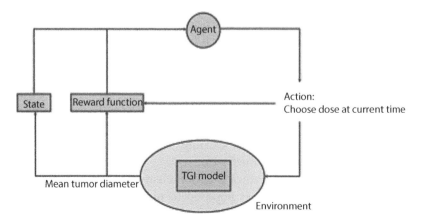

Figure 15.4 Proposed reinforcement learning model.

This paper outlines a reinforcement learning technique for determining the best treatment plan for chemotherapy patients. As the underlying model, Q-learning is also utilized in this research. They create a number of dosages for a specific duration for each action space from which an agent may select. Dosage cycles are only started at a frequency chosen by professionals. The transition states are computed at the end of each cycle. The reward function is the average reduction in tumor diameter. Using simulations, evaluation of the clinical trials is done. Although the exact method of construction of these simulations is unclear, it appears that pathological and statistical data are frequently included.

For each set of parameters they changed, they ran three fundamental experiments to examine the impact of increasingly limited reward systems under various agent flexibilities. Both PCV and TMZ were employed in one set of the simulated studies. Two sets of experiments were conducted for both treatments. In the patient-based experiments, the agent could independently treat different simulated patients. On the other hand, in the trial-based experiments, the agent had to administer the same dose to all simulated patients at each time step. In a separate series of tests, the agent had the flexibility to administer fixed unit doses of TMZ or PCV, as well as doses at concentrations of 25%, 50%, 75%, and 100% at each time step. Throughout the TMZ and PCV studies, the agent also had the option to refrain from administering a dose at any given time step [15].

15.3.3   Dynamic Treatment Recommendation

To significantly improve actual clinical results, dynamic treatment recommendation systems built on sizable electronic health records (EHRs) are necessary. This project teaches a complete treatment strategy using actor criticism approaches and supervised RL. Supervised Reinforcement Learning along with Recurrent Neural Network (SRL-RNN), a collaborative learning framework, is created in this study. The SRL-RNN specifically uses an off-policy actor–critic design to manage complicated interactions among various drugs, diseases, and personal traits. In contrast to RL and Supervised Actor Critique (SAC), with the help of SRL-RNN, information may be transferred from the assessment signal to the indication signal using an end-to-end strategy for learning. Supervised learning and reinforcement learning are combined.

This study is noteworthy since it uses the entire MIMIC-III dataset to attempt to treat all patients, as opposed to only a subset. These factors include time-series variables as well as static variables. Each treatment plan separates time-series data into several units, which are each set to

24 hours because the MIMIC-III prescription frequency is the median. To get around the issue of partial observability, various techniques are employed, including the actor, which offers prescriptions depending on the patient's condition; the critic network, which assesses the usefulness of these medications to encourage or discourage them; and the LSTM (long short-term memory). These three important components make up the majority of this setup shown in Figure 15.5. There are 1,000 precise drugs, or 180 different pharmacological kinds, in the action space. In this study, DTR is modeled as a determinate policy made up of action space and a Markov decision process with a limited number of steps (MDP). POP-20, Basic-LSTM (BL), Reward-LSTM (RL), Dueling Double Deep Q learning (D3Q), Supervised Dueling Double Deep Q (SD3Q), Supervised Actor Critique, and LEAP [16] are some of the comparative techniques employed.

Both when receiving static treatment and when receiving dynamic treatment configuration, and in all accepted baselines, the suggested model SRL-RNN performs better. The following are the causes: (1) In contrast to LEAP and Popularity-20 (POP-20), SRL-RNN views a treatment referral as sequential decision-making with medical considerations practice and uses assessment signals to determine the best course of action. (2) In contrast to D3Q, SRL-RNN views a doctor's prescription as supervisory information to establish an effective policy and uses the off-policy actor–critic structure to manage complex relationships between drugs, illnesses, and individuals.

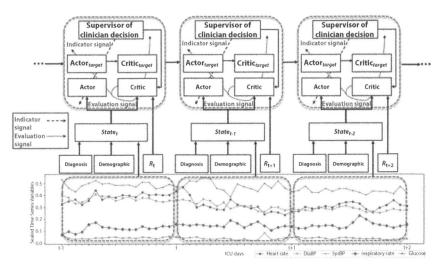

Figure 15.5 Framework for supervised reinforcement learning using a recurrent neural network.

It has been shown that the MIMIC-III technique not only improves drug recommendations but also lowers projected hospital mortality by 4.4% using quantitative research and qualitative case studies. The authors use a projected in-hospital death rate to evaluate their techniques [17].

15.3.4 Dynamic Therapy Regimes Using Data from the Medical Registry

This study is notable since it does not focus on a certain human type but instead tries to create a foundation for a number of dynamic therapy regimes. This study uses observational medical data to propose the first instance of a deep reinforcement learning model to forecast the most efficient dynamic treatment regimens. This approach can mimic the intricacy of various illnesses' progression and available treatments more accurately than current reinforcement learning techniques. Its objective is to offer data-driven, individualized decision-making recommendations to physicians and patients. In order to forecast the anticipated expert behaviors, a supervised training step is included in the deep reinforcement learning approach that has been described. Deep reinforcement learning is also used to determine Dynamic Treatment Regimes' long-term value function.

Supervised learning is the first line of defense against a typical negative effect of donating bone marrow in graft versus host disease (GVHD), where the donor's immune cells assault the patient's cells. Because of guided learning, patients can anticipate a number of professional therapies that may be appropriate for them. The allocation of the first conditioning regimen and GVHD prophylaxis during the transplant time and therapies for small GVHD at 100 days and 6 months and therapies for chronic GVHD at different points of time after the transplant up to a period of 2 years are all predicted using the proposed framework on the basis of time-varying parameters. The agent works to lower the risk of issues during the second stage.

The illustrative data are drawn from the Center for International Blood and Marrow Transplant Research (CIBMTR) registry database, which was established in 1972 with the goal of gathering data on patient outcomes undergoing hematopoietic cell transplantation (HCT). The registry dataset incorporates data from national and international organizations to track the patients' long-term follow-up and to allow comparisons of several decision-stage processes.

The registry dataset incorporates data from national and international organizations to track the patients' long-term follow-up and to allow comparisons of several decision-stage processes as well as to provide a comprehensive deep reinforcement learning structure, as shown in Figure 15.6, that may be applied to challenging situations with broad state and action domains. The two stages are the offline deep neural network construction phase and the online deep Q-learning phase of the deep reinforcement learning technique. The systematic deep reinforcement learning architecture presented in this paper enables sequential decision-making for the first time using data from a medical registry. A well-structured deep reinforcement learning structure is applied to medical observational data with protracted follow-up on an adequate number of patients suffering from a certain set of illnesses.

The system exhibits a high degree of predictive accuracy for treatment choices made by human experts, and theoretically, through reinforcement learning, it has the ability to beat actions taken by observed human experts by maximizing long-term patient outcomes. With its adaptability and versatility for high-dimensional action and diverse decision phases to mimic real-life complexity, this framework improves on existing Dynamic Treatment Regime (DTR) learning approaches. In order to explain a wider action and state space, the expanding reinforcement learning paradigm in statistical literature is adjusted in this context, and a discrete-time model

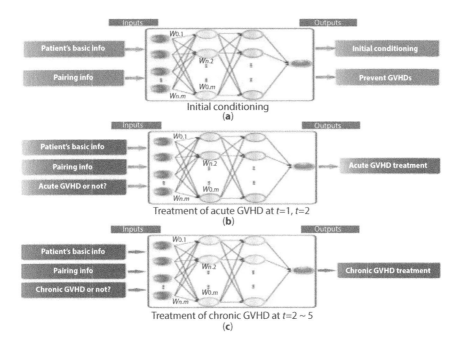

Figure 15.6 Supervised learning frameworks, (a) initial conditioning, (b) treatment of acute GVHD at t=1, t=2, (c) treatment of chronic GVHD at t=2 ~ 5.

is produced, utilizing the registry data gathering approach [18]. The Figure 15.6 shows the visualization of the suggested supervised learning structure, where (a) initial conditioning, (b) treatment of acute GVHD at t=1, t=2, (c) treatment of chronic GVHD at t=2 ~ 5.

15.3.5   Encouraging Physical Activity in Diabetes Patients

Those with type 2 diabetes are known to benefit as a result of regular physical exercise. However, the majority of them are sedentary. Through constant monitoring, communication, and individualized feedback, smartphones open up new opportunities for assisting people in sticking to their physical activity objectives. Twenty-seven sedentary patients with type 2 diabetes were provided with a smartphone-based pedometer and a personalized physical activity program. They also received regular SMS texts encouraging them to engage in physical activity, with messages sent daily to weekly. The messages were customized in order to improve each participant's adherence to the activity plan using a Reinforcement Learning (RL) algorithm. Both a constant message-sending technique and weekly reminders were compared to the RL algorithm. In contrast to patients in the control group who did not receive signals generated by the RL algorithm, the results show individuals who did improve their activity and walking pace. Patients allocated to the RL algorithm group experienced greater levels of blood glucose reductions when compared to control policies, and longer involvement resulted in better blood glucose level reductions. The learning system improved over time at spotting the messages that would inspire consumers to work out.

A smartphone application was created to collect data on patients' physical activity while running in the background of their devices. This information was sent to a central server. Every morning, a reinforcement learning (RL) algorithm determined which SMS message had the best chance of increasing the patient's physical activity the next day. The chosen message was then delivered to the patient. To produce personalized recommendations, the RL algorithm considered user historical activity, expected activity, demographics, and message history. The trial included 27 patients, all of whom successfully installed the smartphone app and provided information for almost a week.

In this study, a system was created to take advantage of the smartphone's capabilities for communication and constant monitoring in order to investigate a novel strategy for raising adherence. A pilot study looked at feedback messages' effect sent directly to patients' mobile phones using a computerized smartphone application that tracked physical activity goals. To produce a highly customized reminder service, the system used reinforcement learning to identify the most helpful input for every person in a specific circumstance. The significance of individualizing messages was brought home by the analysis of various feedback responses, the impact of message sequences, and the contextual circumstances. It was discovered that patients were not happy with getting weekly reminders to engage in physical activity and that these reminders did not increase engagement. On the other hand, boosting both the amount of time spent walking and the velocity of walking was accomplished by altering the messages in accordance with the activity carried out as indicated by the learning algorithm. In fact, the RL algorithm learned to order messages in a way that maximizes effectiveness [19].

15.3.6   Diagnosis Utilizing Medical Images

The value of a medical imaging examination in the context of producing a proper diagnosis is determined by the quality of acquisition of images and interpretation of images. The advancement of radiology has profited greatly from the past century's achievements in image detection and technology. On the other hand, it is easy to make mistakes when trying to interpret a visual. More than 70% of lawsuits that have been filed against medical imaging providers are the result of missed diagnoses [20]. The majority, or 60% of errors, are perceptive errors that result in missed diagnoses [21]. According to several studies, there is a 75% chance of getting a cancer diagnosis that is falsely negative. Due to the high rate of diagnosis errors brought on by incorrect interpretation of images related to medical and the current advancement in automated approaches to support specialists in this error-prone duty, new technologies and processes have been established to assist specialists in making accurate diagnoses. The goal of this chapter was to evaluate the effectiveness of a reinforcement learning strategy for identifying lesions in medical photographs. The approach can be used to identify between benign and malignant lung nodules, as demonstrated by the results of this investigation. A collection of 3D geometric measurements taken from lung lesions will be used in our study.

Medical imaging has undergone a revolution that has allowed for less intrusive, quicker, and more accurate operations. By enabling us to model the patient's physiological functioning and dysfunctions using cutting-edge algorithms and techniques, imaging tools may significantly improve our ability to perceive and understand the biophysical world. The anatomical morphology can be scanned using a variety of imaging techniques. These consist of computed tomography (CT), ultrasound, radiography, and magnetic resonance imaging (MRI). On the other hand, working modalities are employed to examine the functioning of the tissues. They covered scintigraphy, single photon emission computer tomography (SPECT), functional magnetic resonance imaging (fMRI) in this course, and positron emission tomography (PET). A lot of research has been done on how reinforcement learning is proliferating in many application sectors. Some claim that with the right application of such technology, computer-based systems can more easily be integrated into the healthcare setting, creating chances to support and enhance the work of medical professionals, thereby increasing the effectiveness and standard of medical treatment.

The creation of tools that support and aid the expert in their diagnostic job is one of the main application areas for reinforcement learning. Fakih and Das (2006) [22] created a learning-based approach that improved a fair assessment of diagnostic performance based on patient data and suggested test(s). The development of a thorough performance measurement takes into consideration test costs, test-related morbidity and mortality, and diagnosis turnaround times. The diagnostic power of the tests is also taken into account in the performance management. Data mining, utility theory, Markov decision processes (MDP), and rough set theory are all used in the method. From medical databases, rules conveying diagnostic information are extracted using the rough set theory. Several non-homogeneous performance measurements are combined into a single, cost-based measurement using utility theory. An MDP model, in conjunction with an RL algorithm, is utilized to assist in the identification of effective testing procedures. This technique is applied to address a specific diagnostic challenge related to a single lung nodule (SPN). The results obtained from this approach, as visualized in Figure 15.7, are compared with the four other testing techniques. The Fernandes Figueira Institute and the Pedro Ernesto University Hospital, both located in Rio de Janeiro, provided the images for a project focused on developing computer-aided diagnosis (CAD) tools. Thirty-nine nodules in all, 29 of which were benign and 10 of which were malignant, were gathered from diverse genuine patients.

This paper examines one specific application for categorizing lung nodules in-depth, as well as a current literature overview regarding the application of reinforcement learning in medical applications involving image processing. The 3D geometric nodular characteristics were employed as the classification criterion in the reinforcement learning application for lung nodules that was discussed. Although the results are preliminary, they show that the reinforcement learning classifier may successfully distinguish between benign and malignant lung nodules based on CT scans by making utilization of the lesions' geometry. The opportunity to invest more substantially in this area of research arises from the fact that this machine-learning technology is not often employed with medical images.

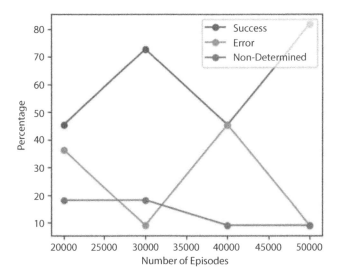

Figure 15.7 Success rate of model.

However, there are certain well-known problems with this application that require more study. We should investigate strategies for cutting training time while maintaining learning effectiveness. Moreover, the tests must be improved to yield more conclusive outcomes and to facilitate comparison with other classifiers [23].

15.3.7   Clinical Research for Non-Small Cell Lung Cancer

In the treatment of advanced metastatic stage IIIB/IV non-small cell lung cancer (NSCLC), multiple lines of therapy are commonly employed. A flexible reinforcement learning technique is suggested to find the best individualized treatment options for individuals with advanced NSCLC who have not previously received systemic therapy through a carefully planned clinical trial. Finding the best time to start second-line therapy, whether on the spot or following induction therapy, is a crucial goal in this difficult undertaking, in order to achieve the longest overall survival time. The selection of the most effective medications for both initial and second-line treatments is based on prognostic characteristics. In the clinical reinforcement study, the Q-learning technique is employed to train an optimal treatment regimen using patient data. To estimate the Q-function using time-indexed variables, a modified version of support vector regression (SVR) capable of handling censored data is utilized. A simulation study showcases the ability of this technique to efficiently and accurately derive the best treatment plans for two stages of treatment, without prior knowledge of how the treatment would actually function. Furthermore, they show that even accounting for individual differences in NSCLC, the design consistently chooses the best first time for second-line therapy.

The foundation of this study is built upon the Q-learning reinforcement learning technique, aiming to optimize the average survival time of patients by considering prognosis information, past treatment decisions, and optimal scheduling. The concept of clinical reinforcement trials utilizing Q-learning was initially introduced by Zhao *et al.* [24] as a means to determine the most effective treatment plans for conditions such as incurable cancers. To address the challenge of irreversible illness status and the existence of multiple treatment options, this approach integrates dynamic therapy regimes within counterfactual paradigms and sequential multiple assignments randomized trials [25]. This therapeutic strategy aligns with the broader concept of personalized medicine.

The clinical reinforcement trial designed for non-small cell lung cancer (NSCLC) incorporates appropriate randomization of patients across various medications in both first-line and second-line therapies. Additionally, the trial includes randomization of the initiation period for second-line treatment. The trial's framework, shown in Figure 15.8, allows for the assessment of optimal personalized treatment plans under hypothetical scenarios [25]. Moreover, reinforcement learning is applied to analyze the gathered data. It is recommended that upon completion of this study, a subsequent confirmatory phase III trial be conducted to determine the

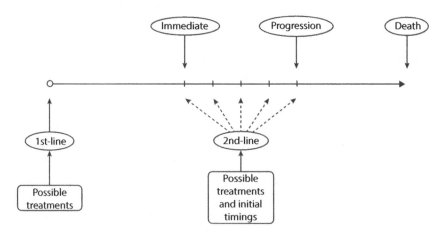

Figure 15.8 A trial treatment strategy and therapy choices for advanced NSCLC.

most effective form of personalized therapy compared to standard care and/or established alternative options [26].

15.3.8   Segmentation of Transrectal Ultrasound Images

One of the most extensively utilized technologies for diagnosis and therapy is ultrasound imaging. The use of ultrasonic imaging in clinical settings is relatively common among the various medical image modalities. However, due to difficulties such as low picture, Ultrasound images are notoriously difficult to segment due to contrast, noise, and missing or blurry edges. One significant use for transrectal ultrasound (TRUS) images is estimating the position and volume of the prostate, as manual segmentation is a time-consuming and labor-intensive operation for this reason.

In this, they described a unique reinforcement learning-based prostate segmentation algorithm, outlined in Figure 15.9, in transrectal ultrasound images. This method is employed to extract the prostate and determine which local values are appropriate for sub-images. There is an offline phase in which photographs and manually split copies of these pictures serve as the foundation for the reinforcement learning agent's learning. The reinforcement agent receives rewards or penalties for objectively exploring and using the solution space. The agent has now acquired knowledge that will be kept in the Q-matrix. With this information in hand, the agent can next use fresh input pictures to extract a rough representation of the prostate. TRUS picture slices are used to implement the proposed method. The dataset consists 60 photos of different prostate shapes. Eight manually segmented photos from the dataset were utilized to instruct the agent offline, and the agent received instruction. The TRUS photos are 468,356 pixels in size.

The RL agent was trained using normal Q-learning on the eight ground-truth photos. This work sought, by displaying some basic results, to introduce the idea of reinforcement learning through the use of TRUS photo segmentation. The original image was broken up into multiple smaller ones.

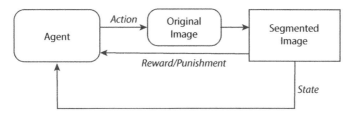

Figure 15.9 Model used in proposed approach.

The agent then altered the environment for each of the ground-truth photographs by altering the threshold level and the length of the structuring element for every sub-image (the quality of the segmented sections).

With all of these samples, the agent should be able to gather the essential information and complete the Q-matrix following the offline stage. Given this knowledge, the agent took actions that resulted in the highest possible rewards for each of the sub-images. Afterwards, it went into internet mode, where the new photographs were displayed. Because it makes use of past input image knowledge, this technique is especially useful when the object and background features of the input images are quite similar. Based on its understanding, the agent may select the proper parameter values for these photos. The suggested method can be used for single-object segmentation tasks, such as prostate segmentation in TRUS pictures [27].

15.3.9   Personalized Control of Glycemia in Septic Patients

Effective glucose management is crucial in critical care, but there is limited research for control of glycemia on individualized optimum approaches. This study aims to develop and customize the best glycemic paths for seriously unwell sepsis patients by determining the desired blood glucose levels as a reference for physicians. To achieve this, a reinforcement learning strategy with policy iteration and a sparse autoencoder was utilized to encode patient states and learn the best course of action from the available data. An algorithm was developed to model the connection between actual glucose levels of blood and 90-day mortality rates by predicting the expected outcomes based on observed glycemic trajectories. Implementing the acquired optimal strategy could potentially reduce the expected 90-day mortality rate by around 6.3%, from about 31% to 24.7%. These findings highlight the capability of reinforcement learning, coupled with proper encoding of patient conditions, to generate optimal glycemic trajectories. This approach empowers healthcare professionals to devise personalized strategies for effective glycemic control in patients.

Poor glycemic control, including dysglycemia and significant glycemic fluctuation, is common in critically sick individuals. The association between blood glucose control and outcomes in patients receiving critical care has been studied by Inzucchi [28], Chase et al. [29], Van den Berghe et al. [30], and others. In order to learn the rule for choosing customized optimized glycemic trajectories based on observational studies, this research suggested and examined the reinforcement learning (RL) methodology. Finally, we compared the generated trajectory's predictions to the actual trajectory's following, which followed the best-case scenario.

The study's dataset was the massive, openly accessible Medical Information Mart for Intensive Care III (MIMIC-III) version 1.4 [31]. Between 2001 to 2012, all patients hospitalized to Beth Israel Deaconess Medical Center's (BIDMC) Intensive Care Unit (ICU), a tertiary healthcare center in Boston, Massachusetts, were included in MIMIC-III. Clinical information, such as laboratory and the bedside measures, along with the time of death (at the event of a fatality), as determined by the Social Security Administration database, are all incorporated into the ICU administrative records.

Using retrospective data, the representation learning and RL algorithms are combined to more precisely forecast glycemic objectives. This decision support could potentially reduce the fatality rate of septic patients by helping physicians define more specific treatment goals and whether the dosages determined by the doctor will actually accomplish the desired glucose concentration picked up by the policy. Figure 15.10 gives an overall visualization of the study. Future studies will utilize a constant state space approach, and several assessment methods, including doubly resilient evaluation, and apply the idea to numerous clinical decision-making problems [32].

15.3.10   An AI Structure for Simulating Clinical Decision-Making

Today's healthcare system makes it challenging to make the best treatment decisions over time due to variables like rapidly rising prices and complexity, an increasing number of available treatment options, and more information being streamed in, often missing the front lines. This study intends to solve these kinds of issues by developing a generic-purpose (non-disease-specific) artificial intelligence (AI)/computational framework. This platform may be used as the foundation for clinical AI as well as a testing ground for "doctor-like" artificial intelligence (AI), new healthcare payment methods, etc.

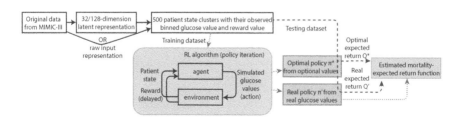

Figure 15.10 Review of the study.

This methodology utilizes Markov decision processing and dynamic decision networks to analyze different sequential decision paths. By learning from clinical data, it generates intricate strategies that take into account the occasionally competing and occasionally cooperative relationships between the various elements of the healthcare system. It can adapt and adjust its actions in partially observable situations as an online agent, continuously updating its understanding of the patient's status of health based on new observations. The effectiveness of this approach was assessed using accurate patient information that was taken from an electronic medical record. The framework of this method is modeled in Figure 15.11.

The outcomes show the viability of this strategy; a framework like this one readily beats the fee-for-service/case-rate healthcare models that are currently used for treatment-as-usual (TAU). The cost per unit of outcome change (CPUC) for AI against TAU was $189 versus $497, demonstrating that AI was less expensive. Despite the cheaper cost, the AI technique resulted in a 30% to 35% improvement in patient outcomes. This advantage might be further strengthened by tweaking a few parameters in the AI model, resulting in a nearly 50% improvement in results at around half the cost. The usage of an AI simulation framework enables optimal decision-making even in ambiguous and complex scenarios, provided the concerns are carefully designed and framed. A future study is described, emphasizing new research avenues and the incorporation of machine learning algorithms in personalized medicine. Clinical data were acquired from Centerstone's electronic health record (EHR), which has about 961 patients who participated in the client-directed outcome-informed (CDOI) pilot study in 2010 and participants in the ongoing post-pilot phase evaluation of CDOI. Results, treatment

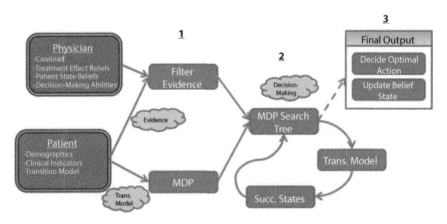

Figure 15.11 Framework overview.

Table 15.1 Overview of all the application.

S. no.	Application	Year	Dataset	Technology used	Pros	Cons
3.1	Treatment of sepsis with deep reinforcement	2017	Multiparameter Intelligent Monitoring in Intensive Care (MIMIC-III v1.4) database [31]	Dueling Double-Deep Q Networks	1. Determining the best course of treatment for sepsis patients 2. Continuous state-space modeling discovered comprehensible regulations that could enhance the treatment for clinical care	Not yet able to examine learned policies on a patient-by-patient basis.
3.2	Chemotherapy and Clinical Trial Dosing Regimen Selection.	2018	(Ribba *et al.*, 2012).	Double-Deep Q Networks, TGI model environment	1. By evaluating different trade-offs between potential harm brought on by activities and outcome indicators associated with the state, the framework learns about new TMZ and PCV dosing regimens 2.The use of action-derived rewards affords greater learning flexibility	In areas where there is no known or suspected relationship between reward and state representation or behaviors, action-derived penalties are intrinsically less helpful.

(Continued)

Table 15.1 Overview of all the application. (*Continued*)

S. no.	Application	Year	Dataset	Technology used	Pros	Cons
3.3	Dynamic treatment recommendation	2018	MIMIC-III v1.4 [31]	Supervised Reinforcement Learning with RNN (Recurrent Neural Network)	1. To discover optimal dynamic treatments and solve the POMDP problem, the SRL-RNN model blends off-policy actor–critic architecture with RNN. 2. Extensive testing on real-world EHR datasets demonstrates that SRL-RNN reduces predicted hospital mortality by up to 4.4% and improves drug prescription.	Understanding and analyzing the suggested model's decision-making process may be more complex than with simpler models.

(*Continued*)

Table 15.1 Overview of all the application. (*Continued*)

S. no.	Application	Year	Dataset	Technology used	Pros	Cons
3.4	Dynamic Therapy Regimes Using Data from the Medical Registry	2017	Center for International Blood and Marrow Transplant Research (CIBMTR) registry database	Deep Reinforcement Learning-based Dynamic Treatment Regimes Framework	1. Deep RL framework proposed employing information from the medical registry, for data-driven sequential decision-making. 2. Existing DTR learning methods were extended for flexible and adaptable high-dimensional actions.	This work is based on offline learning with observational dataset
3.5	Encouraging Physical Activity in Diabetes Patients	2016	26-week research from a tertiary hospital's Endocrinology and Diabetes outpatient clinic	Reinforcement Learning	1. The application of on-policy learning allows the agent to change its policy in real time to the current environment conditions. 2. This technology not only allows for a prediction tool, but it also allows for personalized care.	This is a small-scale study

(Continued)

Table 15.1 Overview of all the application. (*Continued*)

S. no.	Application	Year	Dataset	Technology used	Pros	Cons
3.6	Diagnosis Utilizing Medical Images	2008	Images from Fernandes Figueira Institute and Pedro Ernesto University Hospital in Rio de Janeiro were used for CAD tool development, consisting of 39 nodules (29 benign and 10 malignant) from real patients.	Reinforcement Learning (Q-Learning)	1. Encouraging preliminary results. 2. Reinforcement learning can help distinguish benign from malignant lung nodules based on shape characteristics.	-Need to shorten training phase while maintaining quality. -Improve tests for more definitive results. -Dataset size is too small for conclusive findings.
3.7	Clinical Research for Non-Small Cell Lung Cancer	2011	Simulated Data Generated	Q-Learning Framework + Support Vector Regression	1. Successful incorporation of Q-learning with ε-SVR-C for tailored treatment regimens. 2. Potential application of algorithms to design trials for different diseases.	Complexity of Q-functions in clinical applications for parametric regression.

(*Continued*)

Table 15.1 Overview of all the application. (*Continued*)

S. no.	Application	Year	Dataset	Technology used	Pros	Cons
3.8	Segmentation of transrectal ultrasound images	2008	Random Transrectal Ultrasound images	Reinforcement Learning	1. Potential for method to become the primary segmentation approach. 2. Can be trained with a limited number of samples.	Lacks evaluation using medical imaging quality measures. Basic implementation with room for improvement.
3.9	Personalized Control of Glycemic in Septic Patients	2017	MIMIC-III v1.4 [31]	Sparse Autoencoder + Reinforcement Learning	1. Assists clinicians in optimizing real-time treatment strategy. 2. Enables wider applications in personalized clinical decision-making.	Current evaluation methods may be insufficient, requiring the use of doubly robust evaluation.
3.10	An AI Structure for simulating clinical decision-making	2013	Client-Directed Outcome-Informed (CDOI) pilot study at Centerstone	MDP/POMDP simulation framework	1. Modeling improves complex treatment decisions. 2. Outperforms current healthcare models.	-Unrealistic modeling of health state transitions. -Ethical concerns regarding clinician reimbursement based on performance information.

details, demographic data, and other clinical markers were all included in this clinical information. Major clinical depression was the most common diagnosis among the 5807 participants in our sample, and a large percentage of them (approximately 65%) also had chronic physical illnesses such as diabetes, hypertension, cardiovascular disease, and chronic pain. The Table 15.1 shows a comprehensive inventory of ten applications, providing insights into the years when they were created, the datasets they harnessed, and a comprehensive assessment of their individual merits and limitations.

The research described here adds up to an evolving body of evidence suggesting that modeling may be a more effective method for handling such complex treatment decisions than relying just on intuition. There is also a chance to develop this framework into a technical foundation for delivering individualized medication. Such a strategy offers genuine chances to address the core healthcare issues of our day and could play a significant part in improving human performance [33].

15.4   Challenges

In healthcare, RL has the potential to improve clinical decision-making, personalize treatments, and optimize healthcare operations. Yet, in order to use RL in healthcare, here are some problems that need to be fixed. Despite its early success, RL in healthcare still needs to overcome a variety of significant but manageable challenges prior to being extensively deployed in clinical settings. Moving a reinforcement learning (RL) agent from a simulation or training environment to the actual world is one issue. Another is upgrading or altering the algorithm since the agent can only interpret the rewards and consequences as feedback. One of the biggest challenges of RL in healthcare is data availability and quality. RL algorithms require large amounts of data to learn from, and high-quality data are necessary for producing findings that are accurate. On the contrary, healthcare data are often fragmented, incomplete, and of variable quality. Additionally, there are concerns around patient privacy and data security, which can make it difficult to access and use healthcare data for machine learning purposes. The most effective way for RL agents to learn is through on-the-job training, even though utilizing real patients to train RL algorithms is not the most moral course of action (so to speak). This calls for them to train in realistic environments utilizing historical data of a particular medicine, which is typically hard to get because of a number of factors, such as HIPAA compliance and concerns about personal health information (PHI). The approach to overcoming this difficulty is to use synthetic data generated from existing healthcare data. Synthetic data can be used to train RL algorithms without compromising patient privacy or data security. Additionally, synthetic data can be generated to reflect a wide range of healthcare scenarios, which can improve the generalizability of RL algorithms. Another potential solution is to use federated learning, which allows multiple healthcare organizations to collaborate on the development of machine learning algorithms without sharing patient data directly.

Another challenge of RL in healthcare is safety and ethical concerns. Healthcare is a highly regulated industry, and the potential consequences of errors in decision-making can be severe. It is important to ensure that RL algorithms are safe and ethical and that they do not exacerbate existing healthcare disparities. For example, an RL algorithm that is trained with data that are biased against certain populations may produce biased recommendations or treatments. To address these concerns, there needs to be a focus on developing transparent and interpretable RL algorithms that can be easily understood and validated by clinicians and patients. Additionally, there needs to be a focus on ensuring that RL algorithms are developed and trained on diverse and representative datasets that do not perpetuate healthcare disparities.

One approach to addressing ethical concerns is to involve patients in the development of RL algorithms. Patients can provide input on the types of data that are collected and the types of decisions that are made based on that data. Additionally, patients can help identify potential biases in the data and the algorithms and can provide feedback on the transparency and interpretability of the algorithms. Partial observability is an additional challenge. Despite the fact that RL agents frequently account for the entire state of simulated settings, the human body is much more complex and dynamic than even the most complex simulations or historical datasets. The fact that blood pressure, temperature, and other measures only provide a partial picture of the environment means that RL agents frequently struggle to fully understand it in clinical settings. Interpretability and transparency are also important challenges in RL in healthcare. RL algorithms can be highly complex, and it can be difficult to interpret the reasoning behind their decisions. In healthcare, it is critical to be able to understand and explain the factors that are driving clinical decision-making. This is important not only for clinicians but also for patients who need to understand why they are receiving a certain treatment or recommendation.

To address this challenge, there needs to be a focus on developing interpretability and explain ability techniques that can help clinicians and patients understand the reasoning behind RL recommendations and treatments. This may involve developing visualization techniques, as well as developing methods for tracing the decision-making process of RL algorithms. It can be difficult to design a compensation scheme that balances long-term benefits with short-term (but potentially misleading) improvements, even if RL agents for the healthcare business are designed with long-term success in mind. On the other hand, merely providing one reward at the end leads to a very drawn-out procedure without any intermediate input for the agent (survival or death).

Clinical validation is another challenge of RL in healthcare. RL algorithms need to be validated in clinical settings to ensure that they are effective and safe in practice. This can be challenging due to the complexity of healthcare environments and the need for rigorous clinical testing. Additionally, there may be concerns about the generalizability of RL algorithms, meaning that they may perform well in one clinical setting but not in another. To address this challenge, there needs to be a focus on developing methods for rigorous clinical testing and validation of RL algorithms. This may involve partnering with clinical researchers and healthcare organizations to test RL algorithms in real-world settings, as well as developing methods for validating the safety and efficacy of RL algorithms prior to deployment. Finally, integrating RL algorithms with existing healthcare systems is a significant challenge. RL algorithms need to be integrated with the aid of clinical decision support systems, electronic medical records (EHRs), and other IT solutions for healthcare.

15.5   Conclusion

Reinforcement learning is a reliable and proven method for making optimal decisions in complex tasks that require decision-making with monetary gratification, high-dimensional, noisy input, and complex nonlinear dynamics. Yet, characterizing the condition and action space, along with training and assessing policies using observational data, might be difficult tasks when using this method in the healthcare industry. Another significant challenge is developing reward functions, which must be solved. Ultimately, even if reinforcement learning offers a technically and statistically solid answer for the best possible decision-making in a variety of difficult tasks, its application in healthcare is complicated due to various factors. Designing effective reward functions in healthcare systems is vital to ensure that the healthcare system's goals are achieved without any unexpected outcomes, such as over-treatment or disregarding patient preferences. However, this can be challenging since healthcare outcomes like morbidity and mortality are often delayed, requiring careful balancing of both short-term and long-term results. To develop the most effective reward functions, it is important to consider aspects such as state and action representations, evaluation methods, and the design of reward functions. In the field of healthcare decision-making, it may be necessary to develop new algorithms and methodologies that can address the unique challenges present in this domain. Reinforcement learning, as a technique, has the potential to revolutionize healthcare decision-making. Nonetheless, more work needs to be done before it can be fully realized in healthcare. Overall, it is essential to adapt reinforcement learning to healthcare decision-making to achieve its full potential in this field.

References

1. Cunningham, P., Cord, M., Delany, S.J., Supervised learning, in: *Machine Learning Techniques for Multimedia*. Cognitive technologies, M. Cord, and P. Cunningham (Eds.), Springer, Berlin, Heidelberg, 2008, https:// doi. org/10.1007/978-3-540-75171-7_2.
2. Sutton, R. S. and Barto, A. G., *Reinforcement learning: An introduction* (2nd ed.), The MIT Press, pp. 2–5, 2018.
3. Nemati, S., Ghassemi, M.M., Clifford, G.D., Optimal medication dosing from suboptimal clinical examples: A deep reinforcement learning approach, in: *38th Annual International Conference of IEEE Engineering in Medicine and Biology Society*, pp. 2978–2981, 2016.
4. Parbhoo, S. *et al.*, Combining kernel and model based learning for HIV therapy selection. *AMIA Summits Transl. Sci. Proc.*, 239–248, 2017 Jul 26; PMID: 28815137; PMCID: PMC5543338.
5. Raghu, A. *et al.*, Continuous state-space models for optimal sepsis treatment: A deep reinforcement learning approach, in *Proceedings of the 2nd Machine Learning for Healthcare Conference, Proceedings of Machine Learning Research*, vol. 68, pp. 147–163, 2017.
6. Zhou, Z. *et al.*, How do tumor cytogenetics inform cancer treatments? Dynamic risk stratification and precision medicine using multi-armed bandits, June 17, 2019. Available at SSRN: https://ssrn.com/abstract=3405082 or http://dx.doi.org/10.2139/ssrn.3405082.
7. Wiering, M. and Otterlo, M., *Reinforcement learning: State-of-the-art*, Springer, pp. 3–42, 2012.
8. Shani, G., Pineau, J., Kaplow, R., A survey of point-based POMDP solvers. *Auton. Agents Multi-Agent Syst.*, 27, 1–51, 2013, https://doi.org/10.1007/ s10458-012-9200-2.
9. Naeem, M., Rizvi, S.T.H., Coronato, A., A gentle introduction to reinforcement learning and its application in different fields. *IEEE Access*, 8, 209320–209344, 2020, doi: 10.1109/ACCESS.2020.3038605.

10. Zhu, P., Li, X., Poupart, P., On improving deep reinforcement learning for POMDPs. ArXiv, abs/1704.07978, 2017, n. pag. [Online]. Available: https://arxiv.org/abs/1704.07978.

11. Wei, J., Dolan, J.M., Snider, J.M., Litkouhi, B., A point-based MDP for robust single-lane autonomous driving behavior under uncertainties, *2011 IEEE International Conference on Robotics and Automation*, Shanghai, China, pp. 2586–2592, 2011, doi: 10.1109/ICRA.2011.5979587.

12. Kaelbling, L.P., Littman, M.L., Moore, A.W., Reinforcement learning: A survey. *J. Artif. Intell. Res.*, 4, 237–285, 1996, https://doi.org/10.1613/jair.301.

13. Raghu, A., *et al.*, Deep reinforcement learning for sepsis treatment. ArXiv, abs/1711.09602, 2017, n. pag. [Online]. Available: https://arxiv.org/abs/1711.09602.

14. Sutton, R.S. and Barto, A.G., *Reinforcement learning: An introduction* (1st ed.), MIT Press, pp. 7–9, 1998.

15. Yauney, G. and Shah, P., Reinforcement learning with action-derived rewards for chemotherapy and clinical trial dosing regimen selection, in: *Machine Learning for Healthcare Conference*, 2018, November, PMLR, pp. 161–226.

16. Zhang, Y., Chen, R., Tang, J., Stewart, W.F., Sun, J., LEAP: Learning to prescribe effective and safe treatment combinations for multimorbidity, in: *KDD*, pp. 1315–1324, 2017.

17. Wang, L., Zhang, W., He, X., Zha, H., Supervised reinforcement learning with recurrent neural network for dynamic treatment recommendation, in: *Proceedings of the 24th ACM SIGKDD International Conference on Knowledge Discovery & Data Mining*, 2018, July, pp. 2447–2456.

18. Liu, Y., Logan, B., Liu, N., Xu, Z., Tang, J., Wang, Y., Deep reinforcement learning for dynamic treatment regimens on medical registry data, in: *2017 IEEE International Conference on Healthcare Informatics (ICHI)*, 2017, August, IEEE, pp. 380–385.

19. Hochberg, I., *et al.*, Encouraging physical activity in patients with diabetes: Intervention using a reinforcement learning system. *J. Med. Int. Res.*, 19, 2016, [Online]. Available: https://www.jmir.org/2017/12/e418/.

20. Berlin, L. and Berlin, J., Malpractice and radiologists in cook county, IL: Trends in 20 years of litigation. *AJR Am. J. Roentgenol.*, 165, 781 788, 1995.

21. Renfrew, D.L., Franken, E.A., Berbaum, K.S., Jr *et al.*, Error in radiology: Classification and lessons in 182 cases presented at a problem case conference. *Radiology*, 183, 145 150, 1992.

22. Fakih, S.J. and Das, T.K., LEAD: A methodology for learning efficient approaches to medical diagnosis. *IEEE Trans. Inf. Technol. Biomed.*, 10, 2, 220–228, April 2006, doi: 10.1109/TITB.2005.855538.

23. Netto, S.M.B., Leite, V.R.C., Silva, A.C., de Paiva, A.C., de Almeida Neto, A., Application on reinforcement learning for diagnosis based on medical image, in: *Reinforcement Learning*, p. 379, 2008.

24. Zhao, Y., Kosorok, M.R., Zeng, D., Reinforcement learning design for cancer clinical trials. *Stat. Med.*, 28, 3294–3315, 2009.

25. Murphy, S.A., An experimental design for the development of adaptive treatment strategies. *Stat. Med.*, 24, 1455–1481, 2005a.

26. Zhao, Y., Zeng, D., Socinski, M.A., Kosorok, M.R., Reinforcement learning strategies for clinical trials in nonsmall cell lung cancer. *Biometrics*, 67, 4, 1422–1433, 2011.

27. Sahba, F., Tizhoosh, H.R., Salama, M.M., Application of reinforcement learning for segmentation of transrectal ultrasound images. *BMC Med. Imaging*, 8, 1–10, 2008.

28. Inzucchi, S.E., Management of hyperglycaemia in the hospital setting. *New Engl. J. Med.*, 355, 18, 1903–1911, 2006.

29. Chase, J.G., Shaw, G., Le Compte, A., Lonergan, T., Willacy, M., Wong, X.-W., Lin, J., Lotz, T., Lee, D., Hann, C., Implementation and evaluation of the SPRINT protocol for tight glycaemic control in critically ill patients: A clinical practice change. *Crit. Care*, 12, 2, R49, 2008.

30. Van den Berghe, G., Wouters, P., Weekers, F., Verwaest, C., Bruyninckx, F., Schetz, M., Vlasselaers, D., Ferdinande, P., Lauwers, P., Bouillon, R., Intensive insulin therapy in critically ill patients. *New Engl. J. Med.*, 345, 19, 1359–1367, 2001.

31. Johnson, A.E.W., Pollard, T.J., Shen, L., Lehman, L.-W.H., Feng, M., Ghassemi, M., Moody, B., Szolovits, P., Anthony Celi, L., Mark, R.G., MIMIC-, I.I.I., A freely accessible critical care database. *Sci. Data*, 4, 160035, 122, 2016.

32. Weng, W.H., *et al.*, Representation and reinforcement learning for personalized glycemic control in septic patients. ArXiv, abs/1712.00654, 2017, n. pag. [Online]. Available: https://arxiv.org/abs/1712.00654.

33. Bennett, C.C. and Hauser, K., Artificial intelligence framework for simulating clinical decision-making: A Markov decision process approach. *Artif. Intell. Med.*, 57, 1, 9–19, 2013.

16

Cognitive Computing in Smart Cities and Healthcare

Dave Mahadevprasad V.[1†], Ondippili Rudhra[2†] and Sanjeev Kumar Singh[2*]

[1]Maharaja Krishnakumarsinhji Bhavnagar University, Bhavnagar, Gujarat, India
[2]Department of Bioinformatics, Alagappa University, Karaikudi, Tamil Nadu, India

Abstract

Human intelligence, also known as human cognition, is a perception of the surrounding environment or world that includes aspects of memory, thinking, and learning. Besides, it is one type of process in which data are received through senses and processed. The 21st century has unending sources of data that the human brain cannot handle, so we need tools to analyze this massive amount of data and process it for our own advantage. Today, we have tools like Artificial Intelligence (AI) and Machine Learning (ML) models that, when properly trained, can perform analysis akin to that of a human. Healthcare is currently one of the largest industries, and it generates a lot of data every day from research, medical procedures, automated diagnosis, and so on. AI and ML are already being used to analyze data and provide a better healthcare outcome. Additionally, AI and ML can be used to support local city governance by enhancing transportation, energy efficiency, pollution prevention, and city safety and to ensure order in cities. This book chapter describes the use of AI and ML models in healthcare and building a smart city.

Keywords: Healthcare, internet, computer, cognitive computing, machine learning algorithms, artificial intelligence and reinforcement learning

Corresponding author: sksingh@alagappauniversity.ac.in
[†]First two authors contributed equally to this work

Elakkiya, R. and Subramaniyaswamy V. (eds.) Cognitive Analytics and Reinforcement Learning: Theories, Techniques and Applications, (317–364) © 2024 Scrivener Publishing LLC

16.1   Introduction

Artificial intelligence (AI) is a branch of computer science that has been around for more than 50 years. It also includes mathematical concepts, statistics, probability, and computer science (Figure 16.1). Its objective is to perform specific tasks by mimicking or copying human cognition and intelligence. The first research on AI, or artificial intelligence, began around the time of World War II, but in 2010, it faced a steep curve in its development because of new developments in programming languages as well as new and powerful computers. Speech recognition, machine learning, and language processing are the right, precise, and fundamental applications of AI [1]. Between 1940 and 1960, there was a strong sense of technological progress and a desire to understand how to merge the operation of machines and living people. The Second World War acted as a stimulant for these developments. Norbert Wiener, a pioneer in the subject of cybernetics, saw the necessity to combine mathematical theory, electronics, and automation in order to build "a whole theory of control and communication, both in people and machines". Warren McCulloch and Walter Pitts created the first mathematical and computer model of the biological neuron (formal neuron) as early as 1943 [2]. The average person, businessperson, or any other non-computer man thinks that AI is just one component, but AI requires more specialized hardware and software along with the machine learning algorithms to create a more trustworthy and work-focused AI [3].

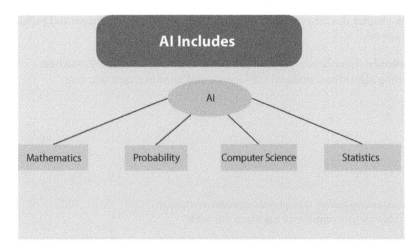

Figure 16.1 Different types of artificial intelligence and prospects.

There are many different programming languages, but only a select few, like Python and R, have the ability to design, train, and deploy a powerful AI (Figure 16.2). AI processes data by taking it as an input, analyzing it, and identifying relationships and particular patterns. It makes predictions for the future using these specific patterns. For instance, a straightforward food image recognition tool is designed to recognize specific types of food and tag them. In order to train it, we must feed it millions of different food images. The tool will then analyze the training data, identify a pattern, and distinguish between the various food images [4, 5].

Algorithms are the fundamental set of rules that enable artificial intelligence to take data as input, process it, and then perform an action (Figure 16.3). Step-by-step work will assist AI in making an accurate analysis of the data, processing it, and producing a consistent and reliable output. AI can assist us in a variety of ways, such as by analyzing large amounts of data that can be used by businesses to track their monthly accounts. Scan the documents to ensure that all of the fields are correctly filled in. For example, if we talk about virology research or finding similar genes for particular diseases, AI can help us to find a similar gene sequence. AI can also be used in research fields such as plant research [6] because there are large amounts of data that are exported from these fields. As a result, the AI can find a similar pattern in all of these large amounts of data and give us conclusive results. Artificial Intelligence may help us in a wide range of capacities, from providing suggestions in response to our Google search queries to assisting businesses in gaining business insights, to locating suitable

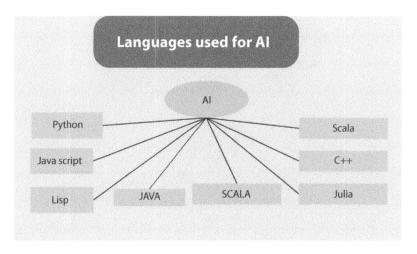

Figure 16.2 The most commonly used programming languages in artificial language.

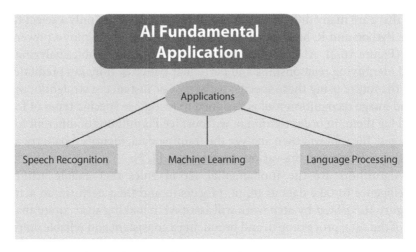

Figure 16.3 Top three applications of artificial intelligence.

recommendations for the route we take to reach our destinations, and to suggesting a suitable hotel. Today, artificial intelligence (AI) is used in every imaginable field, from businesses to schools, from jobs to vacations.

a) How It Is Used in Today's World

Like the examples given above, it is used basically in every aspect of our life (Figure 16.4). For example, Google uses machine learning algorithms to give us proper suggestions and answers to our query. However, to have

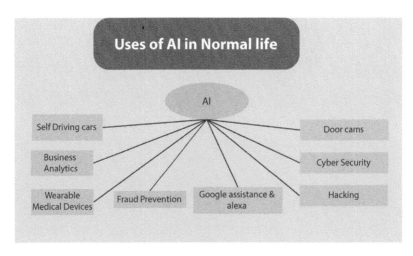

Figure 16.4 The importance of artificial intelligence in everyday life.

a more precise answer to this question, there are mainly two types of AI, namely, face and speech recognition programs, that can be seen in our mobile phones. Today's smartphone has a facial lock feature that can only be unlocked by the specific person's face. In simple terms, it is an image classifier program that finds specific patterns in the face that is shown and it compares the face with the previous face image that was inputted. Other examples are Google Assistant, Alexa by Amazon, and Siri in Apple phones, which are speech recognition programs that are activated when we speak certain words, and after the specific words are formed into a query, it recognizes the speech and finds the answer to that specific query [7]. Nowadays, AI is also being used in houses to do certain tasks like detecting who is out there on the front door, managing the temperature in the house, turning on the lights when you enter the room, turning off the lights when necessary, or cutting off the power in case of overvoltage [8].

Currently, it is also being used in cybersecurity. As we all know, at present, many things are being supported by computers, from schools to businesses, and we all need computers for certain tasks. However, this is also a threat because of the hackers, but nowadays, AI is also being used in cybersecurity in which it can find malicious files and viruses. It can also detect if certain IP addresses are a threat to businesses. It can also be used in ethical hacking. Not all hackers are bad, though, such as when Barnaby jack said "*Sometimes you have to demo a threat to spark a solution*"; thus, these people can also help businesses and governments to patch the security flow or, in other words, critical vulnerability in their critical systems, which, if breached, can be a big threat. Thus, how do hackers use AI in their field of work? AI can help hackers to launch specific attacks in specific patterns, which can also be large scale. Another type is embedded AI or AI that works with hardware. The most famous example of embedded AI is tesla or autonomous cars. These cars use AI to drive their way in. There are also factories that use AI-powered robots for manufacturing products and sort the products for their quality [9].

Nowadays, AI-powered agriculture can calculate the need of the land itself and inform the farmer the amount of land that needs fertilizer and it can also create a system or protocol that can be helpful for the farmer to fulfill the needs of the farming land. It can also be used for cutting unnecessary plants that grow around the farm land by identifying it. It can also be used in disaster management systems to calculate the effectiveness as well as for weather prediction for ease in people mobilization if things go wrong. AI can also be helpful in transportation such as in airplanes, e.g., flight path optimization to automate flight safety components. From flight maintenance to flight scheduling, it can help in many ways. In cars, AI is used to predict possible crashes, and it can also detect bike riders, cyclists, or people walking so there will be a decrease in traffic accidents.

In the rapidly urbanizing landscape of the 21st century, cities around the world are evolving into complex ecosystems with interconnected networks of people, infrastructure, and technology. As these urban centers expand, so do the challenges they face—ranging from traffic congestion and resource management to environmental sustainability and public safety. In response, the concept of smart cities has emerged, redefining urban living by utilizing machine learning (ML) and artificial intelligence (AI) to design more effective, adaptable, and sustainable urban landscapes.

Smart cities can be envisioned as living organisms, where data serve as the lifeblood that courses through the veins of urban infrastructure. The idea is not merely to adopt technology for the sake of modernity, but to utilize it strategically to enhance the quality of life for citizens. AI and ML, two cutting-edge technological advancements, have emerged as indispensable tools in this quest for smarter urban living.

AI, the science of creating intelligent machines capable of mimicking human cognitive functions, has evolved from a concept of science fiction into a practical reality. Machine learning, a branch of artificial intelligence, aims to make it possible for computers to learn from data and develop over time without explicit programming. These innovations give computers the ability to analyze enormous amounts of data, spot trends, and make wise decisions—a capability that precisely satisfies the requirements of smart cities.

16.2 Machine Learning Inventions and Its Applications

The term "machine learning," also known as "ML," was first used by Arthur Samuel, a pioneer in the field of AI. In essence, machine learning increases the capacity of computers to learn and automate processes without explicitly programming them to perform specific tasks. He defined ML as "An area of research that enables computers to learn without being explicitly programmed". This indicates that machine learning gives a program or a machine the ability to learn from a task and get better. AI includes a component called machine learning, which draws insights from data and statistics. It is basically a discipline that gives the machine the power to learn from the data that were given as input or it can also learn from the past experiences or patterns and to give a result to the specific tasks. Machine learning techniques help computers to work independently.

Machine learning models are fed with the data as input and they learn from it, grow and develop themselves for the future specified work or tasks. It generally uses information from data by using some algorithms that are in common terms are just set of rules and they find specific patterns and just learn from it. They do not rely on any predefined processes or equations they use data directly to learn. Machine learning is present in every aspect of our lives, including trading, self-driving cars, surveillance systems, and self-scoring credit (Figure 16.5). Additionally, it can be used to detect mutations in proteins, RNA, or DNA, which is very beneficial for biologists. Likewise, it may be employed in medical imaging to identify body part abnormalities and aid in their treatment by doctors. Lastly, it can be applied to the discovery of new drugs. Nowadays, machine learning is also being used in home securities like motion detectors or motion tracking (Figure 16.5).

There are numerous kinds of machine learning, including reinforcement learning, unsupervised machine learning, and semi-supervised machine learning, which we discuss about later in depth. We can benefit from machine learning in many ways, such as by personalizing care or spotting fraudulent transactions. To measure glucose level, SpO2, and pulse rate, personal assistants can now be embedded in medical devices or, more specifically, wearable medical devices. As it can identify weaknesses in network systems and simulate pen-testing protocols to find weak spots or vulnerabilities in servers or systems, machine learning is also used in cybersecurity and hacking. Besides that, hackers use machine learning

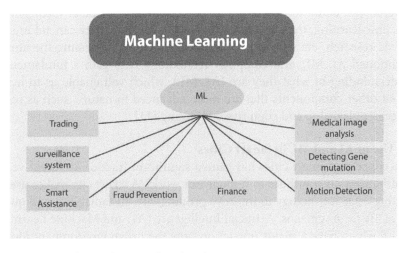

Figure 16.5 Modern application of machine learning.

models to create phishing emails and to get around spam filters. It is also very helpful when attempting to guess passwords. Hackers are also using machine learning models to learn about their targets' defenses, such as which firewall they are using, weak points, or vulnerabilities, as well as traffic patterns.

a) Use of AI and ML in Science

AI and machine learning are used in every sphere of our lives and careers, including science. Every day, the scientific field produces enormous amounts of data. Since a regular person cannot keep up with it, AI and ML are used. In the field of medical science, AI and ML are being used to derive scientific insights for each individual patient to personalize their treatment. Critical illnesses like cardiac arrest, myocardial infarction, and cancers are also detected using it. Additionally, X-rays, CT scans, MRIs, EEGs, ECGs, and EMGs are just a few examples of the medical images that may be analyzed with it. which can be very useful for accelerating the diagnosis of diseases and allowing radiologists to take more images. It can also be used to analyze medical images such as X-rays, CT scans, MRIs, EEGs, ECGs, and EMGs, which is very useful for accelerating disease diagnosis and allowing radiologists to take on more cases. Additionally, there are robots that are AI and ML driven that can be useful for tasks that require precision, such as mixing explosive liquids or working with highly contagious viruses or bacteria [10].

After learning the fundamentals of the vast AI and ML fields and how they are used in science and daily life, it is time to concentrate on the one specific aspect of them that is the subject of this chapter. We have already covered the definitions of computers, technology, artificial intelligence, machine learning, their applications in daily life, how they can aid in scientific research, etc. However, the primary goal of discussing the aforementioned AI, ML, and computer technology is to gain a fundamental understanding of what they are (AI, ML), which will enable us to learn about other components that are more advanced in nature, such as reinforcement learning and cognitive computing [11].

b) Use of AI and ML in Smart Cities

The concept of smart cities has gained significant traction in recent years as urbanization continues to reshape our world. A smart city makes use of cutting-edge technology to improve productivity, sustainability, and quality of life for its citizens. Artificial Intelligence (AI) and Machine Learning (ML), which have a major impact on how the urban environment looks, are at the core of this transition. From optimizing traffic flow to managing energy consumption, AI and ML have become the driving force behind the evolution of smart cities.

One of the most significant challenges in urban areas is traffic congestion. AI and ML are revolutionizing the way cities manage traffic flow and optimize transportation systems. Using data from sensors, cameras, and GPS devices, intelligent traffic management systems predict traffic patterns, identify congested areas, and suggest alternative routes. This not only reduces commute times but also minimizes carbon emissions, contributing to a greener environment.

Ride-sharing platforms also utilize AI algorithms to match riders with drivers efficiently. These algorithms consider factors such as distance, traffic conditions, and historical data to provide users with the quickest and most cost-effective transportation options.

Sustainability is a cornerstone of smart cities. AI and ML are instrumental in monitoring and managing energy consumption. Smart grids use AI algorithms to balance energy demand and supply in real time, optimizing energy distribution and reducing wastage. Additionally, machine learning models can predict energy usage patterns, helping utility providers anticipate peak loads and adjust their operations accordingly.

Furthermore, AI-powered systems manage street lighting by adjusting brightness based on real-time conditions. This not only saves energy but also improves safety and reduces light pollution.

Effective waste management is crucial for maintaining a clean and healthy urban environment. AI and ML are enhancing waste collection processes by analyzing historical data and current conditions to optimize collection routes. Sensors embedded in waste bins provide real-time data on fill levels, allowing authorities to schedule pickups only when necessary, reducing operational costs and environmental impact.

AI-driven environmental monitoring systems track air and water quality, noise levels, and other vital parameters. These data help city planners identify pollution sources, assess the impact of policies, and take proactive measures to ensure a sustainable ecosystem for citizens. Also, AI and ML technologies are making cities safer by enabling predictive policing and surveillance. Advanced algorithms analyze historical crime data to locate crime hotspots and anticipate future crimes. Law enforcement agencies can allocate resources more effectively based on these predictions, deterring criminal behavior and enhancing public safety.

Surveillance cameras equipped with facial recognition algorithms aid in locating missing persons and identifying criminals. However, ethical considerations surrounding data privacy and surveillance transparency must be carefully addressed to maintain a balance between security and civil liberties, and smart cities aim to provide seamless services to citizens. AI-powered chatbots and virtual assistants are being employed to handle citizen inquiries, guide them through bureaucratic processes, and provide real-time information about public services. This enhances citizen engagement, saves time, and improves the overall experience of interacting with local authorities.

Moreover, sentiment analysis tools process data from social media and other sources to gauge public opinions on various issues. This information can help city officials understand citizen concerns and preferences, enabling better-informed policy decisions [12].

16.3   What is Reinforcement Learning and Cognitive Computing?

Machine learning includes reinforcement learning, which is used to develop the capacity to act in a particular way under particular environmental conditions. Businesses can benefit greatly from reinforcement learning because it allows for self-learning, the ability to perform specific tasks with accuracy similar to the human brain, and the ability to improve one's performance based on prior experiences and tasks. It is essentially a component of science where the models learn the appropriate behavior in particular environmental circumstances to get the best result or reward. In other words, it is a decision-making process that enables models to decide or act precisely in a given circumstance for the tasks at hand. To put it simply, humans learn through interaction with the environment that surrounds them, including society, other people, and animals. Through interaction with all of these, humans learn to make decisions about what to do in what situations and how to react in those situations in ways that will help them live their lives and realize their dreams. Because humans learn through reinforcement learning, they can perform this action independently. If we must refer to it in scientific terms, it is known as the trial-and-error theory. Edward Lee Thorndike is credited with coming up with the original version of this theory, which is described as "A basic approach to problem-solving marked by iterative, diversified attempts that are pursued until success or until the practitioner quits away", which means that until a living thing receives the desired reward, they learn to take a particular action in a particular situation by repeating tasks that vary based on the results or rewards they receive. Reinforcement learning, a component of machine learning, is similar to the trial-and-error theory in that the model keeps learning from the provided data until it achieves a specific goal or output. In simple terms, the data generated by the machine learning model will be used as input for the model. The reinforcement algorithm analyzes the output for each process and verifies the feedback. which will aid the reinforcement learning model in determining whether the steps taken or the process followed were appropriate or not. In addition, reinforcement learning can be used in various fields like in manufacturing plants, assembly lines, or image processing in the medical field [13].

16.4   Cognitive Computing

Computers are being taught how to imitate humans' capacity for cognition in the field of cognitive computing, which is a subfield of computer science. People can perform specific tasks thanks to their cognitive abilities, similar to how one would react to specific incidents, perceive knowledge of a specific circumstance and respond to it, or comprehend, process, store, and retrieve information when necessary, or, let us say, how to make a decision in a particular circumstance. All of these things are cognitive abilities of humans. Although computers are faster at processing and performing calculations, they will never be as sophisticated as the human brain. Even though they can perform quick calculations, processes, or even faster analyses, computers must acknowledge that their capacity for thought is inferior to that of the human brain, but now comes cognitive computing, which enables computers to mimic and behave like the human brain. Artificial intelligence and machine learning models with cognitive computing power can understand, process, store, and retrieve information when necessary. They can also learn how to react in specific ambiguous situations. To complete the complex tasks, cognitive computing employs self-learning, and data- and pattern-analytic algorithms.

It is essentially a whole system that extracts the best and highest output from the input data that was gathered from various sources. After some time, the model is able to decide for itself on the given task by using trial-and-error learning techniques. This can help computational tasks to be performed more precisely, and even computer systems can identify the best method to complete a task with the least amount of effort and the most desirable results. They are even capable of identifying potential problems in the future and coming up with the ideal solution before the issue even arises.

a) Reinforcement Learning and Cognitive Computing Uses in Science Field

By leveraging their highly specialized decision-making skills, reinforcement learning and cognitive computing systems can be very beneficial in the domains of life science and medical science. The most typical and well-known application of RL and CC systems is the creation of an effective treatment plan for patients who have been afflicted by a disease for an extended period of time. In this instance, the treatment plan, drug dosage, and the deadline for the follow-up visit can all be created by the models. Additionally, they can evaluate and assess the patient and develop a specific treatment plan for the specific abnormalities the patient is dealing with. Medical professionals can profile a condition by gathering information on a patient's medical history and symptoms, processing it to get a clear picture of the condition the patient is dealing with. Additionally, it may make it easier for doctors to monitor each patient individually. For instance, an RL and CC model can aid medical professionals in keeping track of a patient's blood pressure, SpO2, respiratory rate, and other vital signs (ECG, cell count, glucose level, cholesterol level, etc.) while they are admitted to the intensive care unit. Therefore, these models can notify the doctors when the situation worsens or improves.

It can also be applied to research, where it can reduce costs and generate more solid evidence. Take drug discovery as an example. Additionally, the fields of genomics, proteomics, and other bioinformatics can make use of reinforcement learning and cognitive computing systems to carry out complex tasks that would take a human a very long time to complete, such as gene expression profiling, data mining, junction prediction, or evaluation of P-P interactions, but can be carried out more quickly and accurately by RL and CC systems. Reinforcement learning and cognitive computing systems can also be used in the fields of genomics, proteomics, and other bioinformatics to perform complex tasks that would take a human a long time to complete, such as gene expression profiling, data mining, junction prediction, or evaluation of protein and protein interactions, but can be completed quickly and with greater accuracy by RL and CC systems. Furthermore, some researchers have used CNN algorithms to recognize some fluorescent protein in microscopically captured images of the yeast using the RL and CC system. To better understand the conditions that are or may be present, some researchers also utilize RL and CC systems in neurological diagnosis techniques like EEG and EMG.

Now, we have all seen what reinforcement learning is, what cognitive computing is, how it is being used in our daily life, how scientists and doctors are using reinforcement learning and cognitive computing to solve some of problems related to environments and human health, how doctors are using reinforcement learning to diagnose and treat a long-term illness, and how scientists are using this tool in their research. However, a new question arises: how to create a reinforcement learning or cognitive computing model and which programming languages one should know to create an RL and a CC model. Thus, to date, we have discussed what and how the above-mentioned things are working and are being used in today's world, but now we are going to dive into the technical and scientific world to know more about how to design a reinforcement learning model and how to implement it. We will also see some of the examples of RL models in-depth to understand the workflow and algorithms to gain a deeper knowledge of reinforcement learning. Furthermore, we will do more work in python and R and we will also see the examples that were written in python and R because both languages we are going to talk about are very easy to understand, which will help us to better understand the workflow and steps of the algorithms.

b) Reinforcement Learning and Cognitive Computing Uses in Smart Cities

RL and cognitive computing may greatly contribute to the optimization of many systems and procedures in the overall setting of smart cities.

1. **Traffic Management:** RL algorithms can optimize traffic signal timings by learning from real-time traffic data. The agent interacts with the environment (traffic flow) and learns to adjust signal timings to minimize congestion and travel times.
2. **Energy Management:** Smart grids can employ RL to balance energy demand and supply, deciding when to store or release energy based on real-time consumption patterns and pricing.
3. **Waste Collection:** RL can optimize waste collection routes by learning the most efficient paths based on sensor data that indicate fill levels in waste bins.
4. **Building Energy Efficiency:** RL algorithms can control heating, cooling, and lighting systems in buildings based on occupancy patterns, optimizing energy consumption while maintaining user comfort.
5. **Autonomous Vehicles:** RL is used to train self-driving cars to make complex decisions on the road, learning from simulated or real-world scenarios.

Reinforcement learning can optimize traffic signal timing by learning from traffic data generated by cities and can adjust signal time to avoid congestion and reduce travel time.

Cognitive Computing Applications

Cognitive computing refers to systems that simulate human thought processes, enabling them to understand, reason, learn, and interact in more natural ways. These systems process huge amounts of unstructured data and offer insights that closely resemble human cognition using AI techniques like machine learning, natural language processing, and pattern recognition.

Cognitive computing has a variety of uses in the framework of smart cities, including:

1. **Citizen Interaction:** Cognitive chatbots and virtual assistants can engage with citizens, providing them with information, services, and assistance in a conversational manner.
2. **Data Analysis:** Cognitive systems can analyze and make sense of the massive amounts of data generated in smart cities, providing actionable insights for urban planning and decision-making.
3. **Emergency Response:** Cognitive computing can process real-time data during emergencies, assisting in predicting disaster impacts and recommending efficient response strategies.
4. **Public Services:** Cognitive systems can process citizen feedback from social media and other sources, helping city authorities understand public sentiments and improve service delivery.
5. **Smart Infrastructure Management:** Cognitive systems can predict maintenance needs for critical infrastructure, such as bridges and roads, based on historical data and usage patterns.

In essence, reinforcement learning and cognitive computing are powerful tools in the arsenal of technologies that can transform smart cities into more efficient, responsive, and livable spaces. While reinforcement learning enables autonomous decision-making and optimization, cognitive computing empowers cities to process and make sense of the massive amounts of data they generate, ultimately improving the quality of life and safety for their residents.

16.5   Data Expressed by the Healthcare and Smart Cities

The healthcare sector or institutions are a system whose sole objective is to recognize, address, and prevent human health problems. Three main elements make up the healthcare system: healthcare providers, healthcare delivery settings, and the financial subsystem that underpins the other two. Healthcare professionals can be found in a variety of specialties including dermatology, gynecology, dentistry, medicine, psychiatry, pathology, microbiology, pharmacists, physicians, and surgeons. The healthcare system has many stages where treatment is provided, including primary care, secondary care, and others. At these stages, a lot of data are produced, including demographic information about the patients, their medical histories, assessments of their conditions, treatment regimens, clinical data, and laboratory results. Earlier, all data that were mentioned above were either stored in typed format or handwritten, but now, because of the increase in advancement of computers, these medical data are not converted into digital data or records. Smart cities are brimming with data sources. Sensors and IoT devices that are normally found in devices that are embedded in infrastructure and public spaces collecting data on air quality, temperature, humidity, and traffic flow, and also surveillance cameras that monitor private and public places are providing real-time imagery for security and analysis and are also used by AI and ML. In today's world, social media platforms facilitate citizen engagement and also generate data on public opinions, concerns, and feedbacks in our homes as well as data like energy usage, generating insights into consumption patterns. Also, nowadays, smart cities also have sensory embedded dustbins that can measure how much is the bin filled and they can even separate dry and wet waste for recycling purposes.

AI and ML are also used to reduce travel times and congestion due to optimized traffic management systems. They can reduce utility costs and contribute to a greener environment. Also, they help us to make our streets cleaner and optimize waste collections. They can also help in predicting and managing disasters by helping officials to respond rapidly and decrease damage.

The data generated by smart cities are a treasure trove that, when analyzed with AI and ML, transform urban living into a safer, more efficient, and sustainable experience. From optimizing transportation to enhancing energy management and responding effectively to emergencies, these technologies are essential to making the smart city vision a reality and improving people's lives while also making them much safer. However, responsible data usage, privacy safeguards, and ongoing ethical considerations must remain integral to the smart city evolution.

In the era of rapid urbanization and technological advancements, smart cities have emerged as a solution to address the challenges of urban living. At the core of this transformation are AI and ML, which bring forth a myriad of innovations that make urban life easier and safer. Let us delve into how AI and ML contribute to enhancing lives and ensuring safety in smart cities. The integration of AI and ML technologies in smart cities is revolutionizing urban living by simplifying daily tasks, enhancing safety, and promoting sustainability. The synergistic relationship between data, AI, and ML transforms cities into more efficient, responsive, and secure environments. While embracing these advancements, it is imperative to ensure data privacy, ethical use of AI, and ongoing collaboration among stakeholders to build smart cities that truly enhance lives while prioritizing safety and well-being.

16.6   Use of Computers to Analyze the Data and Predict the Outcome

We can store the common medical data with the aid of computers. The information included patient treatment plans, lab results, evaluations, and numerous other aspects of information pertaining to healthcare. Technology has assisted in the development of tools that can monitor, share, and even raise an alert if the patient is in a serious position using these data. However, this tool also produces a huge amount of data. These types of data, when analyzed, can aid in enhancing healthcare delivery systems. A wide range of data, including ECG, EMG, EEG, X-rays, CT scans, MRIs, laboratory reports, treatment protocols, and other medical data, including blood pressure records, pulse rate records, records of levels of carbohydrates, cholesterol, oxygen levels, and many more, are available in the healthcare system. AI and ML are being used to digitize these records, analyze them, and provide a more actionable insight. Let us now examine how we can create and implement these strategies [14, 15].

16.7   Machine Learning Algorithm

Four groups dominate the categorization of machine learning tasks (Figure 16.6 shows types of machine learning algorithms) [37].

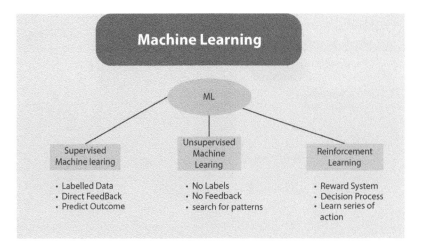

Figure 16.6 Pictorial representation of machine learning algorithm types.

1) Supervised Learning
2) Semi-supervised Learning
3) Unsupervised Learning
4) Reinforcement Learning

a) Supervised Learning

In the case of supervised learning, the machine learning algorithm is provided with training data, which serves as an example of both the intended output from the given input and the conditions or set of rules under which it must be achieved. In a simple way, the model is fed with input data, which is known as training data which will give a desired output or result. The training data are like a guidance system for the algorithm which will take model in a training process and will make predictions, but if the predictions are incorrect, then it will be corrected (Figure 16.7).

This process continues until the goal is achieved with desired precision. In order to identify flaws or defects and fix them in order to provide a desired accurate result, this algorithm can also compare the output with the desired or correct output [16].

Supervised learning technique can be used in:

- **Classification:** to predict the result based on the input or the training data where the result can be in the form of distinct categories. It defines decision boundaries, which includes KNN (K nearest neighbors, Naïve Bayes, and support vector machines).

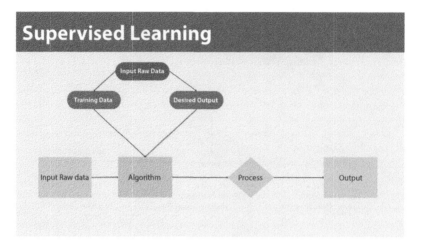

Figure 16.7 Workflow of supervised learning algorithm.

If we take a real-world example, the classification technique can be used for diagnostic purposes.

- **Regression:** regression uses values unlike classification, which includes polynomial regression, support vector machines, linear regression, and neural networks.
- **Forecasting:** also known as time-series forecasting, it is a method in which the prediction is made from past and present data [17].

b) Semi-supervised Learning

Semi-supervised learning is both labeled and unlabeled data as the training data. In these algorithms, models learn to even structure the training data and organize the training data if the training data are unstructured and unorganized [18].

c) Unsupervised Learning

When people talk about self-learning systems, they are actually talking about unsupervised learning. In this method, the algorithm does not have any kind of labeled or structured data. The algorithms will find their own patterns. There will be no assessment for the precision of the result given by the algorithm. The model is created with the interpretation. It finds the structures and hidden patterns in the training data [19].

Unsupervised learning incorporates association, clustering, and reduction (Figure 16.8).

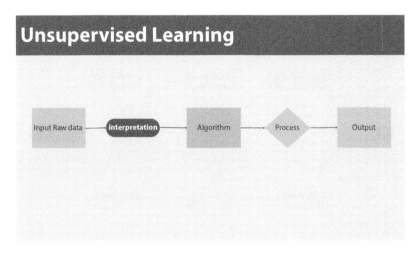

Figure 16.8 Workflow of unsupervised learning algorithm.

- Association: to identify the occurrence of anything present in collection. For example, an algorithm that can predict the percentage of developing lesion or the neurological conditions from stroke or infarction based on the performance or the condition of the heart and cardiorespiratory system.
- Clustering: in this method, clusters are made in which things have the same properties.
- Dimensionality reduction: in this method, data transformation is done from high-dimensional to low-dimensional space with feature selection and extraction. For example, to monitor all the clinical data, lab reports are used to predict the condition of the diseases. It can also be useful for the data of the patient admitted in the ICU. In this situation, the model can predict the state of the patient using the data available.

This method incorporates the k means, neural networks, deep learning, component analysis, and Markov model.

d) Reinforcement Learning

It is a method in which the model has to be more adaptable and more dynamic for the given task. It allows the model to explore the problem, which gives the best feedback possible. RL allows models choose the best outcome based on their current state and also by learning the best behavior

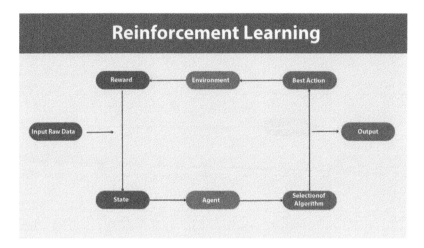

Figure 16.9 Workflow of reinforcement machine learning algorithm.

to maximize the reward and minimize the punishment (Figure 16.9). It is very similar to human intelligence, as humans learn from their experience with their past actions. This method also allows models to learn from their mistakes or their past experience and change their present action to take maximum reward [20].

16.8   How to Perform Machine Learning?

Machine learning methodology methodology (Figure 16.10 reveales basic machine learning model process to solve a task) [38] is simple and it includes the following steps:

- Defining the problem as the learning task,
- Acquiring the data,
- Establishing the learning strategy,
- Implementing it into practice,
- Evaluating the strategy's effectiveness,
- Refining,
- Reporting.

Now, we will see how a machine learning task related to healthcare is performed.

For example, developing a model that will predict the state of the patient, expressed using medical terminologies, will determine the extent to which the patient's disease or prognosis can improve or worsen. In this example, we will talk about developing a model that will analyze the neurological (hemiparesis—paralysis of one side) and musculoskeletal condition (detection of OA—osteoarthritis of knee) of the patient.

a) Specifying the Problem

This step includes understanding what is the problem that we want to solve, what is the procedure that will be used to solve the problem, and how to evaluate the problem. In relation to the previously mentioned condition, the model is being developed to identify the percentage the patient's current condition can develop hemiparesis or OA.

b) Preparing the Data

Models of machine learning learn through the data, so the right data will be very helpful for developing a precise machine learning model. The data have to be in a format that a model can use, for example, clinical and laboratory data of the patient who needs to be diagnosed. It should be cleaned and the data should be fixed.

c) Choosing the Learning Method

After deciding which problem to solve and collecting all the data that a model needs in relation to the task that the model is intended to do, the correct method has to be chosen to develop a precise machine learning model. However, the question is which learning method is good for the

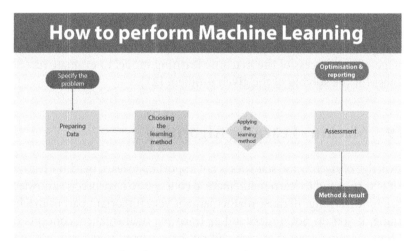

Figure 16.10 Overview of machine learning approaches.

desired machine learning task and how to choose a suitable algorithm for the model?

The choice of algorithm is based on the deadline of the task, size, nature, and quality of the data; the available resources are also a criterion for choosing a correct learning method.

d) Applying the Learning Method

Machine learning is applied in the range of programming languages, but the most famous and used programming languages are python, R, SQL, and MATLAB.

e) Assessing the Method and Results

The machine learning model's task precision is based on the data that were fed to the algorithm. The algorithm does not assess the patient but rather the data that were fed to the model. Thus, the correlation must be done to have confident results.

f) Optimization

To get maximum and precise results, optimization should be done to the model. One can use cross-validation to determine the percentage of confidence in the output that was given by the model.

Algorithms of the machine learning can be optimized by:

- Algorithm tuning
- Training and validation data
- Evaluating a range of methods

g) Reporting

In many institutes, it is very important to report the outcome and the percentage of the success of the machine learning model to evaluate its effectiveness for the solving of the given problem [21].

16.9   Machine Learning Algorithm

Knowledge of algebra and statistics is not important to learn machine learning, but it is very useful to learn mathematical concepts to have precise knowledge of machine learning. Because mathematics is very important part of machine learning, if you do not understand or learn the mathematical concepts of a machine learning algorithm, you will not have proper understanding of the algorithm and its workflow, which will lead to a very limited understanding

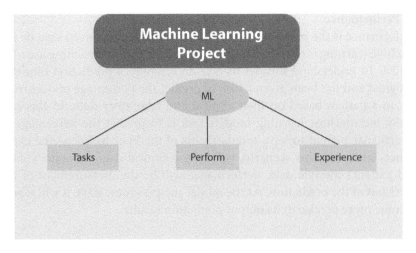

Figure 16.11 Methodology for a typical machine learning approach.

of the algorithm that can also affect the output of the algorithms. By learning the mathematical concepts of the machine learning algorithms, you will learn the basic concepts and workflow of the algorithm that can help you to choose which machine learning algorithm is best suited for the given task.

Now, we will analyze the machine learning algorithm, libraries that were used, and its application in real world. To understand a basic machine learning project, we have to understand its basic component on which a machine learning model works or processes (Figure 16.11 explains about the three main fundamental elements of machine learning project) [38]. Thus, the machine learning project is mainly made up of three things: tasks, experience, and performance.

There are many famous libraries (Figure 16.12 presents the different types of libraries) [38] that will be used basically in every machine learning project such as scikit-learn, numpy, pandas, matplotlib, scarpy, and NLTK. The standard machine learning functions are regression, classification, and clustering, and some common algorithms are k-nearest neighbors, support vector machines, decision trees, and Bayesian network.

a) Tasks
It is basically what is the main function of the machine learning model or it is the main reason a machine learning model is designed.

b) Experience
Experience is the data that was fed to the machine learning model as training data. The amount of data that was fed to the model will determine the machine learning model's precision

c) Performance

Performance is the precision to provide an output of the given data by the machine learning model. It measures the accuracy of the machine learning model. To understand this, let us take an example, a prediction model is designed and has been given a task to predict the percentage of occurring OA in a patient based on their clinical and laboratory data. In this scenario, the machine learning model's task is to predict the percentage of a particular patient to have osteoarthritis of the knee based on the experience, which, in this scenario, will be the clinical data, laboratory data, and patient's lifestyle data; performance will be the measurement of the precision of the prediction. As the model analyzes more data, it will slowly become more precise in its output prediction results.

16.10 Common Libraries for Machine Learning Projects

There are many programming languages that are currently being used in machine learning, but the most famous and common language is python. All the libraries that are used in the python machine learning model is open source, which means that it is available for free and it can be used by anyone. The most common python libraries for machine learning projects are

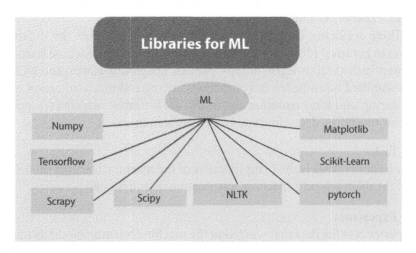

Figure 16.12 The eight best programming libraries for machine learning.

- Numpy: used for functions of mathematics, algebra, and generation of random numbers
- Scipy: used for image processing, signal processing, and ordinary differential equation methods
- Matplotlib: used for data plotting and visualization methods
- Pandas: used for data visualization, manipulation, and aggregation
- Scikit-learn: used for image processing and machine learning techniques like clustering, classification, and regression
- NLTK: used for natural language processing, tokenization, stemming, and classification
- Scrapy: used for data mining and scraping web data
- Tensor flow: designed by Google itself, it gives machine learning models a quick training and multi-layered neural network.
- Keras: also used for building neural networks

16.11   Supervised Learning Algorithm

The main goal is to make a precise functional model. The model is designed to have particular data and function in a specific environment using mathematical algorithms. For example, a model is fed with the data of the blood pressure of the patient and it would predict the chances of cardiac condition's occurrence in the patient. The data being fed to the model can have a wide variety and a large amount like blood pressure value for every week or every day, or even in a specific environment. Training data could be even the food the patient is having or also the primary conditions like kidney disease, narrowing of blood arteries, sleep apnea, or infection or if the patient is having any addiction like alcohol or smoking addiction. These can be input data for the model to predict hypertension in patients.

The simplest form of the model is:
$f(x) = ax + b$, where a and b represent constants, and b denotes the noise in the data, where the peak values a and b are used in such a way that the prediction $f(x)$ is as precise as possible. The model will take clinical data as training data and find patterns in it. However, the precision of the results and output itself has to be correlated with the data and the conditions the patient is in because of some errors that can occur while training the model, like corrupted data, noise, and bias of data.

Table 16.1 Usage of machine learning models in healthcare and the model that was used for the project.

Healthcare usage of machine learning	Machine learning model used	Applied or experiment
Predicting diagnosis	Supervised machine learning Decision tree	Applied
Predicting post stroke pneumonia	Recurrent neural network	Experiment
Chronic diseases prediction using medical notes	Long short-term memory network, convolutional neural network	Experiment
Skin cancer classification with deep neural network	Convolutional neural network	Experiment
Using deep learning to detect pneumonia on chest x-rays	Convolutional neural network	Applied
Breast cancer screening	Convolutional neural network	Experiment
Prediction of diabetic retinopathy	Deep convolutional neural network	Experiment
Using deep belief network for MRI classification for Alzheimer's diseases detection	Deep belief network	Experiment
Predict tacrolimus stable dose in renal transplant recipients	RT	Experiment
Using deep learning off-target prediction in CRISPR-Cas9	Deep convolutional neural network	Applied
Optimized CRISPR guide RNA design for two high-fidelity Cas9 variants by deep learning	Recurrent neural networks	Applied
Developing the homology directed repair (HDR)-mediated nucleotide editing for determining high-efficiency target sites using machine learning	Random forest	Applied

In the supervised learning process, there are primarily two methods

- Classification
- Regression

a) Classification
This approach uses KNN, logistic regression, decision tree, and many more classification techniques.
   Uses of this approach are as follows:

- Diagnosis of patient's condition based on their clinical profile
- Suggesting best treatment protocol for the given conditions, signs, and symptoms
- Risk of recurrence of the diseases
- Assessing the patient's rom is increased or not from the previous assessment of the patient

b) Regression
This approach uses random forest, linear regression, and many more.
   Uses of this approach are as follows:

- Predicting the percentage of developing a cardiac condition based on the blood pressure data of the patient
- Predicting the recurrence of the infection from the blood reports of the patients
- Predicting the percentage of developing diabetes based on the blood glucose reports and eating habits of the patient

16.12   Future of the Healthcare

Having the technology like this can open the door to artificial intelligence-based healthcare where so many things can become a reality (Figure 16.13 illustrates various healthcare sectors where ML can be used) [39]. In the future, your Google Assistant, Alexa, or Siri can schedule your doctor's appointments just by listening to your voice. Your health records or your smartphone will tell you if your health condition is good or bad; you may need to see your doctor based on your blood glucose records. DNA profiling will tell the patients' health condition and the risk of occurrence of particular diseases even while inside the mother's womb (Table 16.1 teaches usage of machine learning model in healthcare) [41]. Combining the patients'

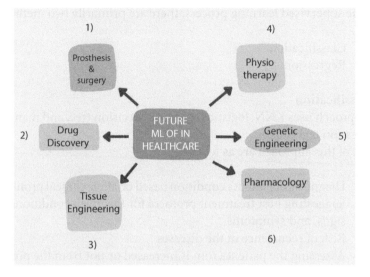

Figure 16.13 Future of machine learning (ML) in healthcare.

data with their devices and clothes, as well as their medical data, assessment records, medical images, MRIs, CT-scans, and X-rays can improve the effectiveness of the diagnosis and treatment protocols.

a) Physiotherapy

A doctor can assess the patient's condition, pain, and other components like muscle strength, range of motion of a particular joint, pulse rate, heart rate, history of illness, and respiratory rate, and diagnose the condition of the patient and develop a precise physiotherapy protocol for the patient for that particular condition.

b) Pharmacology

Personalized medications can now be created with the aid of artificial intelligence and machine learning, increasing their effectiveness while reducing side effects. One can also design personalized blood glucose, cholesterol, and creatinine level stripes that can be used by the patient to monitor their blood glucose, cholesterol, and creatinine levels.

c) Gene Therapy

Using the machine learning with nanotechnology, one can manipulate cells and alter DNA and allow gene editing. Gene editing has the capacity to diagnose, give treatment, and prevent diseases in children born with a genetic condition [22].

d) Tissue Engineering

With the help of artificial intelligence and the increase in technological advancement in healthcare, now medical professionals are not transplanting human organs but printing them with the use of 3D printers. These bio printers are not just printing organs but cells too [23].

e) Drug Discovery

Drug discovery is a very cost-effective field; creating a new drug is very costly and there is a very low percentage of success in this field. Bringing any new drug in a market is very costly too. If a drug trial fails, it can bring very bad consequences to pharmaceutical companies. Because of such consequences, pharmaceutical and research companies cannot afford more errors; it is for this reason they are using artificial intelligence and machine learning models to develop drugs with minimum error possible. AI can be used to develop such compounds that are specific for the diseases that they are going to be used to treat. While giving them all the data of the condition as input, they can design such a compound with higher efficiency and better ways to cure the said condition. Also, this process can even result in drug discovery, thus possibly making the drug available in the market much faster [24, 25].

f) Prosthesis

According to WHO, approximately 30 million people need prosthetic limbs. Technologies like 3D printing, IoT, and cognitive computing may create a very human-like artificial limb. Collecting data like measurements of the patients can be very useful to make a comfortable prosthetic limb, and if it is integrated with sensors and machine learning models, it can work like a normal human limb. Predictive movement algorithms with the combination of human brain and body help control the artificial limb to work like a human limb.

g) Robotic Surgeries and Exoskeletons

Surgery needs precision to be carried out effectively; if robots assist surgeons in surgery, it will improve precision and dexterity. Robotic surgery blurs out the possible percentage of liability, which can be a barrier. Also, because of the decreased production in software and hardware, robotic surgery may become more common in the future. Some surgeries like total knee replacement, dynamic hip replacement, shoulder replacement, and hip replacement surgery are being done with the assistance of robots. Robots can also help patients with disability perform their everyday activity through integrated devices or external appliances. Exoskeletons have become very common in the field of rehabilitation, assisting human mobility under more realistic environments [26].

h) Diagnosis

Artificial intelligence is currently being used in the diagnosis of cardiovascular diseases, diabetes, etc. They can also be used to analyze and assess medical and radiological images like MRIs, CT scans, EEGs, EMGs, ECGs, and X-ray images. With the aid of smart watches, a US-based firm can recognize type 2 diabetes with an accuracy rate of 85% [27–29].

16.13   Development of Model and Its Workflow

The machine learning model's success depends on the steps that are developed (Figure 16.14 explains how ML model is developed) [40] and the assessment of the machine learning model for the given task.

- Prototype phase
- Validating phase
- Deployment phase

The early steps are therefore referred to as the prototype phase, during which the model is supplied with a variety of historical data in order to test and select the optimal machine learning model. The best machine learning model will be tested and confirmed after that. The model is deployed in production for use. After the deployment, the model is evaluated (as per shown in Table 16.2) [42] on performance metrics.

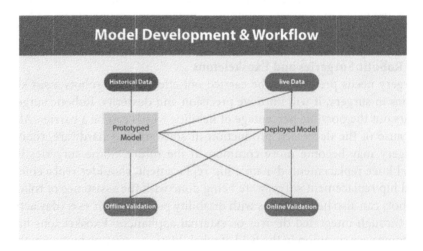

Figure 16.14 Machine learning model development and workflow.

Table 16.2 Evaluation metrics for machine learning models.

Classification	Regression
Accuracy	r squared and adjusted r squared
Confusion matrix	Root mean squared error
Per-class accuracy	Mean absolute error
Logarithmic loss	Percentiles of errors
Area under the curve	RMSE
F-measure	MAPE

16.13.1 Types of Evaluation

- Online evaluation
- Offline evaluation

a) Online Evaluation
Online evaluation is an evaluation of metrics after the deployment of the model. The main point of the online evaluation is that in online evaluation, the training data are different from the data that were used in the past.

b) Offline Evaluation
Offline evaluation is an evaluation of metrics when the model is in training phase, when it is fed with past data, which are stationary. This method includes the hold-back and n-fold cross-validation method.

Joining AI and ML for healthcare can transform life globally. Once shown only in films, new advances in healthcare are now becoming a reality. Gene therapy, tissue engineering, printing human organs, and robotic surgeries have become a reality and will be more precise in the future. Improvement in technology affects both medicine and public's perception on health. The effect of ML, data science, and artificial intelligence in healthcare systems is very noticeable, which will require impartiality.

16.14 Future of Smart Cities

The future of smart cities is poised to be a transformative journey, where AI and ML will serve as the driving forces behind unprecedented advancements in urban living. As these technologies continue to evolve, they will shape how cities operate, innovate, and adapt to the dynamic needs of their residents. Here is a detailed look at the potential future of smart cities in relation to AI and ML [30].

a) Hyper-Personalized Services
AI and ML will enable cities to provide citizens with hyper-personalized services. Smart systems will learn from individual behaviors, preferences, and historical data to offer tailored recommendations for transportation routes, energy usage, entertainment options, and more. This level of customization will enhance citizen satisfaction and make urban life more convenient and enjoyable.

b) Autonomous Infrastructure
Smart cities will embrace autonomous infrastructure, including self-adjusting energy grids, self-repairing roads, and AI-controlled buildings. These systems will use ML algorithms to continuously optimize their operations, leading to efficient resource utilization, reduced maintenance costs, and enhanced resilience against various challenges.

c) Predictive Maintenance and Sustainability
AI-powered predictive maintenance will revolutionize urban infrastructure management. Sensors and data analytics will anticipate equipment failures, enabling proactive repairs and minimizing service disruptions. This approach will extend to sustainability efforts, with ML models predicting energy consumption patterns and guiding conservation measures.

d) Seamless Transportation
Transportation will see remarkable changes with the advent of AI-driven autonomous vehicles and smart traffic management systems. These technologies will not only reduce traffic congestion and emissions but also enhance safety and accessibility for all citizens, including those with mobility challenges.

e) Cognitive Urban Planning
AI-powered cognitive systems will assist city planners in making informed decisions. These systems will analyze vast amounts of data, from citizen feedback to environmental data, to propose optimized urban designs that prioritize efficient land use, green spaces, and sustainable infrastructure [31].

f) Efficient Resource Management
AI and ML will optimize resource management across the board. Smart water management systems will detect leaks and manage distribution

efficiently, reducing waste. Energy grids will balance demand and supply seamlessly, and waste collection routes will be optimized to minimize fuel consumption and pollution.

g) Data-Driven Governance
Governance will become more data-driven and transparent. AI will analyze public sentiment and feedback from social media, assisting policymakers in making informed decisions that reflect the needs and aspirations of citizens.

h) Disaster Preparedness and Response
AI and ML will significantly improve disaster management by analyzing real-time data to predict and manage emergencies. Evacuation plans will be optimized, and first responders will receive real-time insights for more effective disaster relief efforts [32].

i) Enhancing Quality of Life
The overall quality of life in smart cities will be elevated. AI-driven healthcare systems will provide personalized medical insights and telemedicine options. Smart homes will be energy-efficient, secure, and responsive to residents' needs [33].

j) Ethical and Inclusive Implementation
The future of smart cities hinges on ethical AI and ML implementation. Governments and stakeholders will need to address issues of data privacy, algorithm bias, and ensure that technology benefits all citizens, fostering inclusivity and equitable access to smart city resources [34].

16.15   Case Study I

1) Case Study: Machine Learning in Healthcare - Osteoarthritis Assessment, Diagnosis, and Treatment

    a) Introduction: Osteoarthritis (OA) of the knee is a common degenerative joint disease that affects millions of people worldwide. Leveraging machine learning techniques in healthcare can significantly enhance the assessment, diagnosis, and treatment of OA, leading to more accurate and personalized care for patients.

b) **Problem Statement:** Osteoarthritis of the knee poses challenges in early diagnosis, progression tracking, and tailoring effective treatments. The goal is to develop a comprehensive machine learning solution that combines various data sources to assess, diagnose, and recommend personalized treatment strategies for patients with knee OA.

c) **Solution:** Machine learning algorithms can analyze patient data, medical images, and clinical records to predict disease progression, aid in diagnosis, and suggest tailored treatment plans. This case study will focus on a machine learning solution that encompasses different stages of OA management.

d) **Data Collection**
   1. **Patient Medical Records:** Historical medical records provide information about patient demographics, comorbidities, and prior OA treatments.
   2. **Imaging Data:** X-rays, MRI scans, and other imaging modalities capture joint structural changes, which are crucial for assessing OA severity.
   3. **Biomarker Data:** Biomarkers in blood or synovial fluid can offer insights into the disease's biochemical aspects.

e) **Data Processing:** Data from diverse sources are processed and standardized. Feature engineering techniques extract relevant information from images and clinical records to create meaningful inputs for the machine learning models.

f) **Machine Learning Models:**
   1. **Diagnosis and Severity Assessment:** Classification models trained on patient records and imaging data can aid in diagnosing OA and determining its severity.
   2. **Disease Progression Prediction:** Time-series models can predict the progression of OA based on historical imaging data, biomarkers, and patient characteristics.
   3. **Treatment Recommendation:** Recommender systems can suggest personalized treatment plans based on patient demographics, medical history, and disease stage. This can include medications, physical therapy, or surgical options.

g) **Real-Time Decision-Making:** The predictive models continuously analyze incoming patient data and imaging results. They provide real-time insights to clinicians, helping them make informed decisions about diagnosis, treatment, and follow-up care.

h) **Benefits:**

1. **Early Diagnosis:** Machine learning aids in detecting OA at an early stage, enabling prompt intervention and potentially slowing disease progression.

2. **Personalized Treatment:** Tailored treatment recommendations improve patient outcomes and reduce trial and error in finding effective interventions.

3. **Objective Severity Assessment:** Machine learning provides a more objective way to assess disease severity compared to traditional methods.

4. **Patient Engagement:** Personalized treatment plans enhance patient engagement and adherence, leading to better long-term management.

i) **Challenges:**

1. **Data Quality:** Ensuring accurate and reliable data from various sources is crucial for reliable predictions.

2. **Interpretability:** Some machine learning models might lack interpretability, which is important in a medical context for building trust among clinicians.

3. **Ethical Considerations:** Handling patient data raises privacy and ethical concerns. Adhering to data protection regulations is paramount.

4. **Clinical Integration:** Integrating machine learning solutions into clinical workflows and gaining clinician acceptance require careful implementation.

j) **Conclusion:** Machine learning offers transformative possibilities in healthcare, particularly in the assessment, diagnosis, and treatment of osteoarthritis of the knee. By leveraging diverse patient data, medical images, and clinical records, machine learning models can provide accurate diagnosis, predict disease progression, and recommend tailored treatments. This case study demonstrates the potential of machine learning to revolutionize medical practices, improving patient outcomes and contributing to more effective and personalized healthcare strategies.

16.16   Case Study II

1)   Artificial Intelligence's Importance in Renal Disease

In order to enhance medical progress, artificial intelligence (AI), a cutting-edge science technology, has been extensively employed in the disciplines of early disease detection, diagnosis, and treatment. Kidney disease continues to be an international health issue because of the large patient population. The diagnosis and treatment of it still present difficulties. AI has the potential to consider individual circumstances, provide appropriate recommendations, and make significant advancements in the management of kidney disease. Here, we cover recent research on the use of AI in kidney disease alerting systems, diagnostic support, therapeutic guidance, and prognosis assessment. Despite the limited number of studies on AI applications in kidney illness, doctors are aware of the promise of AI in the management of kidney disease, and AI will significantly improve physicians.

Kidney disease also poses a significant public health problem, with common causes caused by diabetes, hypertension, obesity, and aging. The frequency of these diseases is increasing. Worldwide, 750 million people have kidney disease, according to the 2015 Global Diseases, Injuries and Risk Factors Survey. Kidney disease poses a significant burden on society, with a 2017 study showing that the annual cost for patients with stage 3 chronic kidney disease (CKD3) is about $1,205, for those with CKD4 $1,963, and for those with CKD5 $8,035. It turns out there is $34,554 for haemodialysis patients. Early detection and prevention of progression of renal disease to end-stage renal disease is therefore of great importance [35]. AI is the science of computer-simulated thought processes and human behavior, including computer science, psychology, philosophy, and linguistics. In 2016, AlphaGo beat a professional human player 3-0. It was the first computer program to beat a Go world champion, clearly demonstrating the technological potential of AI in this era. The continued rapid growth of computing power, the availability of large datasets, and the development of advanced algorithms over the past two decades have greatly improved machine learning [35].

Electronic medical records (EMRs) provide the large-scale, real-world clinical data that are the basis for developing AI technologies in the clinic. Since it takes a lot of time and care to avoid human errors, as well as the ability to extract deep insights or information, it is challenging for individuals to directly examine these enormous amounts of data. There is little doubt that AI technology outperforms humans in a number of areas. The study of AI with kidney diseases is still in its early stages. According to the available research, the role of AI in kidney illness primarily focuses on four aspects: warning systems, diagnostic support, therapeutic guidance, and prognosis evaluation.

a) Computer-Aided Diagnosis—Diagnostic Assistance (CADD Assistance)

The method known as computer-aided diagnosis (CAD) uses medical imaging and computer image processing to quantify and evaluate the focus's features. It can help physicians quickly and accurately identify and analyze lesions. The role of CAD has been demonstrated in a variety of contexts, particularly in tumors like skin, breast, and lung cancer. Studies on kidney disease are few and mostly focus on pathology and imaging diagnosis.

b) Wearable Dialysis Devices

Dialysis is the primary method of ESRD treatment. High aspirations for the development of wearable artificial kidneys have been inspired by the extensive consequences of dialysis on the patient's life, including certain patients' inability to withstand the hemodynamic instability of intermittent dialysis. Wearable dialysis devices can examine equipment alerts, dialysis parameters, and patient-related data in real time with a real-time feedback reaction. Martin *et al.*, combined AI with regenerative medicine technology to produce wearable dialysis equipment. These devices have little hemodynamic impact, effective toxin removal, and continuous dialysis capabilities (Figure 16.15).

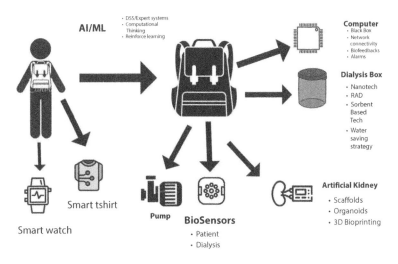

Figure 16.15 Smart kidney uses for end-stage renal disease (ERSD) patients.

A total of 15 ESRD patients have tested the devices. The results showed that dialysis was effective and had no negative side effects. The devices received approval from the American Food and Drug Administration as groundbreaking technology. Nevertheless, the sample size is small; more studies with more patients are required to enhance the model in the future. Another intriguing approach is the wearable artificial kidney, a 5-kg wearable, miniaturized device with a sorbent-based hemodialysis system that is being developed at the University of Washington in the United States. Ten patients were given a wearable artificial kidney for 24 hours as part of an exploratory clinical experiment. The implantable Kidney Assist Device (iRAD), another groundbreaking invention, creates a biohybrid system that can imitate the morphology and function of the kidneys using micromachining techniques (Figure 16.16 demonstrates the role of ML in renal

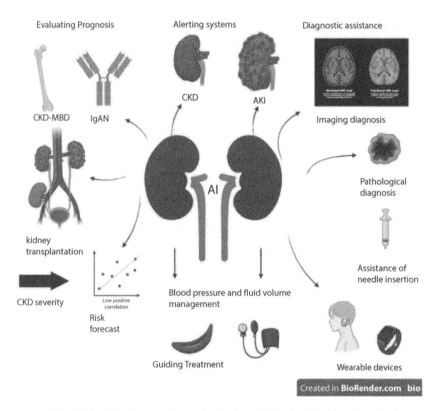

Figure 16.16 Artificial intelligence in renal disorders aids in the identification of patients with chronic kidney disease (CKD), including those with autosomal dominant polycystic kidney disease (ADPKD), diabetic kidney disease, and nephropathy (IgAN).

diseases) [41]. This artificial kidney is a bionic device that incorporates a silicon nanopore membrane and a bioreactor containing living kidney cells in order to concentrate the ultrafiltrate into urine. The bundle is kept in a body-friendly cage and connected to the patient's bladder and circulatory system. Although scaling it up to the clinical situation has been successfully applied in animal models, this work is still under progress. Scientists recently developed MXene Sorbents for urea removal from dialysate, which offer a new possibility for creating a tiny dialysate regeneration system for a wearable artificial kidney, irregular blood and dialysate flow rates as a result of g kinks and variable pump performance [36].

c) Summary
It is clear that AI is widely applied in a variety of domains, not simply nephrology. Take urology as an illustration, where artificial intelligence is utilized to predict genitourinary cancers. We are currently observing the evolution of medical practices, from empirical medicine to evidence-based medicine, intelligent diagnosis, and AI-directed medicine.

Although artificial intelligence (AI) in medicine is still in its infancy, there is no doubt that in the near future, AI will produce prediction algorithms suitable for routine clinical application by exploiting the diversity and complexity of real-world data. These findings go beyond the investigation of the initial targets for data analysis and interpretation, which are potentially expensive, lengthy clinical trials with a small patient population that may eventually supplement or even completely replace real-world data-driven risk assessments. As a result, the discussion of future medical evidence may be fundamentally altered. AI will have a bigger role in clinics in the future since it might boost medical productivity and reduce stress. The present investigations, however, lack significant multicenter studies and are largely retrospective studies. Applying AI to renal problems will require additional research.

16.17   Case Study III

1) Case Study: Machine Learning in Smart Cities—Optimizing Traffic Management

    a) Introduction: Smart cities aim to improve the quality of life of residents by leveraging technology and data to enhance urban services and infrastructure. One critical challenge faced by urban areas is managing traffic congestion efficiently. In this case study, we will explore how machine learning techniques can be applied to optimize traffic management in a smart city environment.

b) **Problem Statement:** Traffic congestion leads to wasted time, increased pollution, and decreased productivity. The goal is to implement a system that uses real-time data from various sources (traffic cameras, sensors, and GPS devices) to predict and manage traffic flow effectively, thereby reducing congestion and improving overall transportation efficiency.

c) **Solution:** Machine learning algorithms can analyze historical and real-time data to predict traffic patterns, congestion, and even suggest optimal routes for commuters. In this case study, we will focus on a machine learning solution using a combination of data sources and predictive models.

d) **Data Collection:**
   1. **Traffic Cameras:** Cameras installed at key intersections capture live traffic images, which are used to monitor congestion and analyze vehicle movement.
   2. **Sensors:** Roadside sensors collect data on vehicle speed, density, and flow. These sensors provide a more granular view of traffic conditions.
   3. **GPS Data:** Anonymous GPS data from vehicles and smartphones offer insights into traffic patterns, route choices, and vehicle speeds.
   4. **Historical Data:** A repository of historical traffic data helps identify trends, seasonal variations, and recurring congestion patterns.

e) **Data Processing:** Data from various sources are processed and aggregated to create a comprehensive traffic picture. Techniques such as data cleaning, normalization, and feature extraction are employed to ensure accurate and relevant inputs for the machine learning models.

f) **Machine Learning Models:**
   1. **Traffic Flow Prediction:** Time-series forecasting models, such as LSTM (Long Short-Term Memory) networks, can predict traffic flow for specific times and locations based on historical data.
   2. **Congestion Prediction:** Classification models like Random Forest or Gradient Boosting can predict whether an area will experience congestion based on current and historical data.

3. **Route Optimization:** Reinforcement learning algorithms can be used to suggest optimal routes for vehicles to minimize travel time and congestion. These models can adapt to real-time changes in traffic conditions.

g) **Real-Time Decision-Making:** The predictive models continuously analyze incoming data from cameras, sensors, and GPS devices. They provide real-time insights into traffic management systems, enabling them to make informed decisions. For instance, traffic lights can be dynamically adjusted based on predicted traffic flows to optimize signal timing.

h) **Benefits:**

1. **Reduced Congestion:** By predicting traffic congestion and suggesting optimal routes, the system helps alleviate congestion, reducing travel times and fuel consumption.

2. **Environmental Impact:** Reduced congestion leads to lower greenhouse gas emissions, contributing to a cleaner environment.

3. **Efficient Resource Allocation:** Traffic management authorities can allocate resources such as traffic police and emergency services more effectively.

4. **Improved Urban Planning:** Data-driven insights into traffic patterns help urban planners design road infrastructure and transportation systems more efficiently.

i) **Challenges:**

1. **Data Quality:** Ensuring the accuracy and reliability of data from various sources can be challenging. Poor-quality data can lead to inaccurate predictions.

2. **Privacy Concerns:** Handling GPS and camera data raises privacy concerns. Proper anonymization and data protection measures must be in place.

3. **Model Adaptation:** Real-time traffic conditions are dynamic and can change rapidly. Models need to adapt quickly to these changes.

4. **Infrastructure:** Setting up the necessary sensors, cameras, and data processing infrastructure can be costly and require ongoing maintenance.

j) **Conclusion:** Machine learning plays a crucial role in transforming smart cities by optimizing traffic management. The integration of various data sources and predictive models empowers traffic management authorities to make informed decisions that improve transportation efficiency, reduce congestion, and enhance the overall urban living experience. This case study demonstrates the potential of machine learning to address complex urban challenges and create more sustainable and livable cities.

16.18   Case Study IV

1) Case Study: Machine Learning in Smart Cities—Waste Management Optimization

a) **Introduction:** Smart cities leverage technology and data-driven solutions to enhance urban living. One significant challenge that urban areas face is efficient waste management. In this case study, we will explore how machine learning can be applied to optimize waste collection and disposal processes in a smart city.

b) **Problem Statement:** Inefficient waste management leads to environmental pollution, increased costs, and operational inefficiencies. The goal is to develop a system that uses machine learning to predict waste generation patterns, optimize collection routes, and reduce the environmental impact of waste disposal.

c) **Solution:** Machine learning algorithms can analyze historical and real-time data to predict waste generation patterns and optimize waste collection routes. This case study will focus on a machine learning solution that combines data sources and predictive models to enhance waste management.

d) **Data Collection:**

1. **IoT Sensors:** Smart waste bins equipped with sensors measure fill levels, temperature, and humidity. These sensors provide real-time data on bin status and environmental conditions.

2. **Historical Data:** A database of historical waste generation data helps identify patterns, seasonality, and fluctuations in waste production.

3. **Weather Data:** External factors such as weather conditions can influence waste generation rates. Integrating weather forecasts into the model improves prediction accuracy.

e) **Data Processing:** Data from various sources are processed and aggregated to create meaningful insights. Data pre-processing techniques such as filtering, normalization, and feature extraction ensure accurate inputs for the machine learning models.

f) **Machine Learning Models:**
1. **Waste Generation Prediction:** Time-series forecasting models, such as ARIMA (AutoRegressive Integrated Moving Average) or LSTM (Long Short-Term Memory) networks, can predict waste generation based on historical data and external factors.
2. **Route Optimization:** Routing algorithms, such as Genetic Algorithms or Ant Colony Optimization, can optimize waste collection routes based on predicted waste generation, bin fill levels, and real-time traffic data.
3. **Bin Monitoring:** Anomaly detection models can identify bins with unusually high or low fill levels, indicating potential overflow or inefficiency.

g) **Real-Time Decision-Making:** The predictive models continuously analyze incoming data from sensors, weather forecasts, and historical records. They provide real-time insights into waste management systems, enabling informed decisions like optimizing collection routes and scheduling pickups.

h) **Benefits:**
1. **Efficient Collection:** Predicting waste generation and optimizing collection routes reduce unnecessary trips and minimize fuel consumption, leading to cost savings.
2. **Reduced Environmental Impact:** Optimized routes and reduced overflow in bins contribute to a cleaner environment and lower greenhouse gas emissions.
3. **Resource Allocation:** Waste management authorities can allocate resources such as collection trucks and personnel more effectively based on predicted demand.
4. **Data-Driven Insights:** The system generates valuable insights into waste generation patterns, allowing for better urban planning and policy decisions.

i) **Challenges:**
1. **Data Accuracy:** Ensuring the accuracy of data from various sensors and sources is essential for reliable predictions.

2. **Model Adaptation:** Rapid changes in waste generation patterns due to events like festivals or holidays require models to adapt quickly.

3. **Integration:** Integrating IoT sensors, data processing systems, and waste management infrastructure requires careful coordination.

4. **Privacy Concerns:** Handling sensor data might raise privacy concerns. Proper anonymization and data security measures are necessary.

j) **Conclusion:** Machine learning offers innovative solutions to transform waste management in smart cities. By predicting waste generation patterns, optimizing collection routes, and enabling informed decision-making, the system enhances operational efficiency, reduces environmental impact, and improves overall urban living conditions. This case study exemplifies how machine learning's application can address complex urban challenges, contributing to more sustainable and livable cities.

16.19   Conclusion

In this chapter, we looked at the Cognitive Analytics and reinforcement learning emerging approaches for a healthcare and smart city's technological advancement. Different types of requirements imposed by healthcare and smart cities development were also examined. Specifically, we highlighted the issue of underutilizing the big data produced by healthcare industry and smart cities, and we proposed various artificial intelligence and machine learning algorithms to address the current challenges and promising future research directions to smart cities and healthcare services growth. Lastly, we came across several problems and existing research orientations for utilizing machine learning algorithms to implement better healthcare and smart city services.

References

1. Schank, R.C., What is AI, anyway? *AI Mag.*, 8, 4, 59–59, 1987.
2. Schank, R.C., Where's the AI? *AI Mag.*, 12, 4, 38–38, 1991.
3. Krige, J. and Barth, K.-H., Introduction: Science, technology, and international affairs. *Osiris*, 21, 1, 1–21, 2006.

4. Barry, A. and Slater, D., Introduction: The technological economy. *Econ. Soc,* 31, 2, 175–193, 2002.

5. Bertelsen, O.W. and Bødker, S., Introduction: Information technology in human activity. *Scand. J. Inf. Syst.,* 12, 1, 3, 2000.

6. Rencilin, C.F. and Rudhra, O., Bioinformatics approaches for high-density linkage mapping in rice research. *Appl. Bioinform. Rice Res.,* 29–54, 2021.

7. Rizomyliotis, I. *et al.,* "How may I help you today?" The use of AI chatbots in small family businesses and the moderating role of customer affective commitment. *J. Bus. Res.,* 153, 329–340, 2022.

8. Giddens, A., *Runaway world: How globalization is reshaping our lives,* Routledge Taylor & Francis, Abingdon, UK, 2003.

9. Chareonwongsak, K., Globalization and technology: How will they change society? *Technol. Soc.,* 24, 3, 191–206, 2002.

10. Pai, F.-Y. and Huang, K.-I., Applying the technology acceptance model to the introduction of healthcare information systems. *Technol. Forecast. Soc. Change,* 78, 4, 650–660, 2011.

11. Ai, W.I., Artificial intelligence (AI) in healthcare and research. Nuffield Council on Bioethics, pp. 1–8, 2018.

12. Marr, B., How is AI used in education–Real world examples of today and a peek into the future, vol. 25, Forbes, Forbes Magazine, New Jersey, U.S., 2018.

13. Barto, A.G. and Sutton, R.S., Reinforcement learning, in: *Handbook of Brain Theory and Neural Networks,* pp. 804–809, 1995.

14. Libai, B. *et al.,* Brave new world? On AI and the management of customer relationships. *J. Interact. Mark.,* 51, 1, 44–56, 2020.

15. Smith, R.G. and Eckroth, J., Building AI applications: Yesterday, today, and tomorrow. *AI Mag.,* 38, 1, 6–22, 2017.

16. Bi, Q., Goodman, K.E., Kaminsky, J., Lessler, J., What is machine learning? A primer for the epidemiologist. *Am. J. Epidemiol.,* 188, 12, 2222–2239, 2019.

17. Cunningham, P., Cord, M., Delany, S.J., Supervised learning, in: *Machine Learning Techniques for Multimedia: Case Studies on Organization and Retrieval,* pp. 21–49, 2008.

18. Abney, S., *Semisupervised learning for computational linguistics,* CRC Press, Boca Raton, Florida, U.S., 2007.

19. Dayan, P., Sahani, M., Deback, G., Unsupervised learning, in: *The MIT Encyclopedia of the Cognitive Sciences,* pp. 857–859, 1999.

20. Fradkov, A.L., Early history of machine learning. *IFAC-PapersOnLine,* 53, 2, 1385–1390, 2020.

21. Brynjolfsson, E. and Mitchell, T., What can machine learning do? Workforce implications. *Science,* 358, 6370, 1530–1534, 2017.

22. Hunt, C. *et al.,* Recent progress of machine learning in gene therapy. *Curr. Gene Ther.,* 22, 2, 132–143, 2022.

23. Guo, J.L., Januszyk, M., Longaker, M.T., Machine learning in tissue engineering. *Tissue Eng. Part A,* 29, 1–2, 2–19, 2023.

24. Chen, Z. *et al.*, Applications of artificial intelligence in drug development using real-world data. *Drug Discov. Today*, 26, 5, 1256–1264, 2021.
25. Vamathevan, J. *et al.*, Applications of machine learning in drug discovery and development. *Nat. Rev. Drug Discov.*, 18, 6, 463–477, 2019.
26. Chang, M. *et al.*, The role of machine learning in spine surgery: The future is now. *Front. Surg.*, 7, 54, 2020.
27. Çelik, Ö., A research on machine learning methods and its applications. *J. Educ. Technol. Online Learn.*, 1, 3, 25–40, 2018.
28. Hamet, P. and Tremblay, J., Artificial intelligence in medicine. *Metabolism*, 69, S36–S40, 2017.
29. Kononenko, I., Machine learning for medical diagnosis: History, state of the art and perspective. *Artif. Intell. Med.*, 23, 1, 89–109, 2001.
30. Ullah, Z., Al-Turjman, F., Mostarda, L., Gagliardi, R., Applications of artificial intelligence and machine learning in smart cities. *Comput. Commun.*, 154, 313–323, 2020.
31. De Las Heras, A., Luque-Sendra, A., Zamora-Polo, F., Machine learning technologies for sustainability in smart cities in the Post-COVID era. *Sustainability*, 12, 22, 9320, 2020, DOI: 10.3390/su12229320.
32. França, R.P., Monteiro, A.C.B., Arthur, R., Iano, Y., An overview of the machine learning applied in smart cities, in: *Smart Cities: A Data Analytics Perspective*, M.A. Khan, F. Algarni, M.T. Quasim (Eds.), pp. 91–111, Springer International Publishing: Cham, U.K., 2021.
33. Sharma, H., Haque, A., Blaabjerg, F., Machine learning in wireless sensor networks for smart cities: A survey. *Electronics*, 10, 9, 1012, 2021, DOI: 10.3390/ electronics10091012.
34. Bhattacharya, S. *et al.*, A review on deep learning for future smart cities. *Internet Technol. Lett.*, 5, 1, e187, 2022.
35. Kelly, B.J., Chevarria, J., O'Sullivan, B., Shorten, G., The potential for artificial intelligence to predict clinical outcomes in patients who have acquired acute kidney injury during the perioperative period. *Perioper. Med. (Lond.)*, 10, 1, 49, 2021.
36. Yao, L. *et al.*, Application of artificial intelligence in renal disease. *Clin. eHealth*, 4, 54–61, 2021.
37. Ayodele, T.O., Types of machine learning algorithms, in: *New Advances in Machine Learning*, vol. 1, p. 1948, 2010.
38. Zhou, Z.H., *Machine learning*, Springer Nature, U.S., 2021.
39. Callahan, A. and Shah, N.H., Machine learning in healthcare, in: *Key Advances in Clinical Informatics*, p. 279291, Academic Press, London, U.K, 2017.
40. Schelter, S., Biessmann, F., Januschowski, T., Salinas, D., Seufert, S., Szarvas, G., On challenges in machine learning model management. *IEEE Data Eng. Bull.*, 41, 5–15, 2018.

41. Nishat, M.M., Faisal, F., Dip, R.R., Nasrullah, S.M., Ahsan, R., Shikder, F., Asif, M.A., Hoque, M.A., A comprehensive analysis on detecting chronic kidney disease by employing machine learning algorithms. *EAI Endorsed Trans. Pervasive Health Technol.*, 7, 29, e1, 2021.

42. Zhou, J., Gandomi, A.H., Chen, F., Holzinger, A., Evaluating the quality of machine learning explanations: A survey on methods and metrics. *Electronics*, 10, 5, 593, 2021.

Index

Printed and bound by CPI Group (UK) Ltd, Croydon, CR0 4YY

27/10/2024

14580132-0003